THERE ARE (NO) STUPID QUESTIONS... IN SCIENCE

THERE ARE (NO) STUPID QUESTIONS

...IN SCIENCE

WRITTEN AND ILLUSTRATED BY
LEAH ELSON

BLACK STONE
PUBLISHING

Copyright © 2023 by Leah Elson
Published in 2023 by Blackstone Publishing
Cover and book design by Kathryn Galloway English
Illustrations by Leah Elson

All rights reserved. This book or any portion
thereof may not be reproduced or used in any manner
whatsoever without the express written permission
of the publisher except for the use of brief quotations
in a book review.

Printed in the United States of America

First edition: 2023
ISBN 979-8-200-86493-5
Science / Essays

Version 1

Blackstone Publishing
31 Mistletoe Rd.
Ashland, OR 97520

www.BlackstonePublishing.com

*To sentient beings of the cosmos,
and the perpetually curious at heart.*

CONTENTS

INTRODUCTION	xi
BIOLOGY	1
HOW DO BACTERIA RESIST ANTIBIOTICS?	3
HOW DOES DNA WORK?	5
DOES CHICKEN NOODLE SOUP ACTUALLY HELP WITH BEING SICK?	7
WHEN CATS PURR, WHAT IS ACTUALLY MAKING THE NOISE?	9
HOW DO VACCINES WORK?	11
WHAT IS THE FUNCTION OF NIPPLES ON MEN?	15
DOES SINGING TO PLANTS REALLY HELP THEM GROW?	17
HOW DOES A VIRUS MUTATE?	19
WHAT HAPPENS TO SPERM AFTER FERTILIZATION?	21
WHY ARE MEN TALLER THAN WOMEN—WHAT PURPOSE DOES IT SERVE?	23
IS NOT HAVING SEX HARMFUL?	25
ARE GMO FOODS BAD FOR YOU?	27
ARE VIRUSES LIVING ORGANISMS?	31
DO MEN HAVE A BIOLOGICAL CLOCK?	33
WHY DOES EVERYONE SAY BEES DYING IS A PROBLEM?	35
ARE THERE ANY DISEASES THAT HAVE JUMPED FROM HUMANS TO ANIMALS?	37
IS HONEY ACTUALLY BEE POOP?	39
IF HUMANS COME FROM MONKEYS, WHY ARE THERE STILL MONKEYS AROUND?	41
WHAT IS THE PURPOSE OF A UVULA?	43
WHAT IS A TARDIGRADE?	45
IS IT POSSIBLE TO CLONE A WOOLLY MAMMOTH?	47
CHEMISTRY	49
WHY DO HOT PEPPERS BURN MY MOUTH EVEN WHEN THEY'RE COLD?	51
HOW DO BAKING POWDER AND BAKING SODA WORK?	53
HOW DOES GLOW-IN-THE-DARK PAINT OR STICKERS WORK?	55
HOW DOES RUBBING ALCOHOL KILL GERMS?	57

WHAT IS FIRE MADE OUT OF?	59
WHAT IS ACID RAIN?	61
WHY DOES ADDING SALT TO A PLUMBING SYSTEM "SOFTEN" THE WATER?	63
WHY ARE SOME ELEMENTS RADIOACTIVE AND OTHERS ARE NOT?	65
WHAT MAKES ICE SLIPPERY?	67
COULD YOU ACTUALLY MAKE A DIAMOND OUT OF COAL?	69
WHY IS CARBON MONOXIDE SO DANGEROUS?	71
HOW DOES FLUORIDE IN TOOTHPASTE HELP PREVENT CAVITIES?	73
HOW DO BATTERIES STORE ELECTRICITY?	75
WHAT IS DRY ICE?	77
WHY DOES IRON RUST, BUT GOLD DOES NOT?	79
WHY DOES HOT, BOILING WATER SHATTER A COLD DRINKING GLASS?	81
WHY DOES COOKING AN EGG YOLK MAKE IT SOLID?	83
WHAT IS BPA IN PLASTIC, AND WHY IS IT HARMFUL?	85
HOW IS THE PERIODIC TABLE ORGANIZED?	87
HOW DO THEY MANUFACTURE FIREWORKS TO BE DIFFERENT COLORS?	89

PHYSICS 91

HOW IS A 735,000-POUND AIRPLANE ABLE TO FLY?	93
WHY DOES ICE FLOAT?	95
WHAT IS QUICKSAND?	97
WHY IS THE SKY BLUE?	99
HOW DO MAGNETS WORK?	101
WHY DON'T HUGE SHIPS SINK, EVEN THOUGH THEY'RE MADE OUT OF METAL?	103
WHAT WOULD HAPPEN IF EARTH STOPPED ROTATING?	105
HOW DO I KNOW THAT THE COLOR GREEN THAT I SEE IS THE SAME COLOR GREEN THAT YOU SEE?	107
HOW ARE ALL THE COLORS OF THE RAINBOW CONTAINED IN WHITE LIGHT?	109
ARE THERE MORE ELEMENTS THAT EXIST THAT WE HAVEN'T DISCOVERED?	111
HOW DO CELL PHONES WORK TO TRANSMIT CALLS?	113
IS TELEPORTATION POSSIBLE?	115
IF I FIRED A BULLET STRAIGHT INTO THE AIR, WOULD IT COME BACK DOWN AT THE SAME SPEED?	117

IS IT TRUE THAT A FEATHER AND A BOWLING BALL
    WILL LAND AT THE SAME TIME IF YOU DROP THEM ON THE MOON?   119

WHAT IS THE DIFFERENCE BETWEEN DIRECT CURRENT (DC)
    AND ALTERNATING CURRENT (AC) POWER SOURCES?   121

HOW DO PHOTONS CARRY COLOR?   123

HOW DO MICROWAVE OVENS HEAT FOOD?   125

WHY ARE NUCLEAR WEAPONS SO POWERFUL?   127

ARE THERE MORE COLORS THAN ROYGBIV?   129

COULD YOU FLY AN AIRPLANE UP INTO SPACE?   131

HUMAN PHYSIOLOGY   133

WHY IS BLOOD RED?   135

WHY IS THE HAIR ON MY ARMS SO SHORT,
    WHILE THE HAIR ON MY HEAD KEEPS GROWING?   137

WHAT IS THE PURPOSE OF PUBIC HAIR?   139

WHY DOES ASPARAGUS MAKE MY PEE SMELL BAD?   141

WHAT CAUSES STUTTERING?   143

WHY DOES MY EYE RANDOMLY TWITCH?   145

WHY DO I SEE RANDOM COLORS WHEN I PUSH ON MY EYES?   147

WHAT IS CANCER?   149

WHICH ORGANS CAN HUMANS LIVE WITHOUT?   151

WHAT IS A HANGOVER?   153

HOW DOES PAIN MEDICINE (ACETAMINOPHEN) WORK?   155

WHAT IS PUS?   157

MY WIFE'S BLOOD TYPE IS O, BUT NEITHER OF
    HER PARENTS HAS TYPE O BLOOD. SHOULD SHE BE CURIOUS?   159

WHY DO WE AGE?   161

HOW DOES THE HUMAN SENSE OF SMELL WORK?   163

WHY DO WE EXHALE CO_2 IF WE INHALE O_2?   165

WHAT IS THE FUNCTION OF THE APPENDIX?   167

IS THERE ANY SCIENCE TO SUGGEST THAT GUT FEELINGS ARE REAL?   169

WHY DO SOME THINGS (HEART, LUNGS, ETC.)
    MOVE ON THEIR OWN, BUT MY ARMS AND LEGS DO NOT?   171

WHY DO HUMANS CRAVE UNHEALTHY FOODS IF THEY'RE SO BAD FOR US— SHOULDN'T WE BE DRAWN TO HEALTHY FOOD?	173
CAN YOU DIE FROM HEARTBREAK?	175
HOW DOES CAFFEINE KEEP YOU AWAKE?	177

SPACE 179

HOW DOES GRAVITY WORK?	181
WHAT DOES SPACE SMELL LIKE?	183
HOW DO WE KNOW THE UNIVERSE IS EXPANDING?	185
WHAT ARE SATURN'S RINGS MADE FROM?	187
WHAT IS THE SUN'S FUEL SOURCE?	189
IF EARTH'S GRAVITY IS PULLING ON THE MOON, WHY DOESN'T IT CRASH INTO US?	191
IS THERE ACTUALLY A HOLE IN THE OZONE LAYER, AND WHAT IS IT FROM?	193
IS IT TRUE THAT ALIENS LIVING SIXTY-FIVE MILLION LIGHT-YEARS AWAY WOULD STILL SEE DINOSAURS IF THEY LOOKED AT EARTH?	195
WHAT THE HELL IS DARK MATTER?	197
IF NO LIGHT CAN ESCAPE BLACK HOLES, HOW DO WE KNOW THAT THEY EXIST?	199
WHY ISN'T PLUTO A PLANET ANYMORE?	201
WHAT ARE ORGANIC MOLECULES, AND WHY ARE WE LOOKING FOR THEM ON MARS?	203
WHY CAN'T ANYTHING TRAVEL FASTER THAN THE SPEED OF LIGHT?	205
CAN MOONS HAVE RINGS OR THEIR OWN LITTLE MOONS?	207
IF THERE WAS A BIG BANG, DO WE KNOW THE LOCATION IN THE UNIVERSE WHERE IT STARTED?	209
IS THE MILKY WAY SIMILAR TO OTHER GALAXIES?	211
HOW MANY GALAXIES ARE IN THE UNIVERSE?	213
IF THERE ARE TRILLIONS AND TRILLIONS OF STARS WITH POTENTIAL PLANETS, WHY HAVEN'T WE DISCOVERED ALIENS YET?	215
WHY IS EARTH THE ONLY PLANET IN OUR SOLAR SYSTEM THAT HARBORS LIFE?	219
WHY DO STARS FLICKER IN THE NIGHT SKY?	221

ACKNOWLEDGMENTS	223
NOTES	225

INTRODUCTION

The idea for *There Are (No) Stupid Questions . . . in Science* was actually conceived six years ago. But to understand its strange, slightly dangerous inception, I'd have to take you back much, *much* further.

You see, dear reader, I've always been a bit of a hambone. When I was growing up, my report cards read the same: "Brilliant student, but very poor citizenship. Talks too much." (However, I'd like the record to show, as I also tried to explain to my parents back then, that "too much" is quite a subjective assessment.) But to be perfectly honest with myself, and you, not much has changed. My superiors in science might likely draft a similar report card for me today if given the opportunity. And this overflow of curious energy brings me to the book you're holding now. This introduction is the chaotic origin story of a scientist you never needed or wanted.

So six years ago, during a live video feed on Facebook, I placed a string of Christmas tree lights in the microwave and fired it up. Why? To confirm a theory that they would illuminate. And also, to make people laugh. Through an electric cascade of sizzles and pops, I explained to my internet onlookers the electromagnetic rationale for why the bulbs lit up when the microwave was running. After the smoke cleared (both the literal and metaphorical kinds), I took my fledgling steps in a journey through science communication.

As the experimentation featured in the live videos became more grandiose, I was urged by my growing group of avid viewers to take the science instruction public. After reconstructing a tiny dual-phase rocket engine model in my bathroom, and accidentally torching a cloth shower curtain to ashes, I decided to move my content to a public platform (in a less flammable capacity). On Instagram I developed my next generation of science outreach, called *60 Seconds of Science*. Under the username @gnarlybygnature, I blazed through easy-to-digest explanations of scientific or medical topics in under sixty seconds. After the initial episodes, viewers began to submit requests for clarification on topics ranging from human disease to the origins of the known universe. With the page's growing popularity, I decided to give full control to my follower base, allowing them to suggest topics and subsequently vote on impending episodes.

There Are (No) Stupid Questions . . . in Science adopts the same premise of that popular series: inform the public by teaching them *exactly* what they want to know. Therefore, the content of this book is made up of 103 explanations to *real* scientific questions, submitted directly to me or harvested from Yahoo! Answers. Topics are separated into five sections: biology, chemistry, physics, human physiology, and space.

Since the dawning of my career in science communication, I have always maintained the same goal: present science in a rapid and digestible way, and demonstrate to any audience that science can be wonderfully silly, occasionally bizarre, but always awe-inspiring.

This book is meant to educate and bridge disparities in science literacy by encouraging any and all questions. Unlike many books, this one welcomes topics from *all* major fields. Being a scientist, I understand implicitly the vastly important and monumentally positive impact that asking questions can have on the human connection to the physical world. Throughout history, insatiable curiosity is what has driven the fundamental discoveries that have shaped our understanding of the world today.

So! In an effort to provide some much-needed clarification to our most humble human quandaries (which, paradoxically, happen to elicit some of the most profound answers), I have decided to embark on this journey. In this book, you will find a collection of real questions from around the world, and the most curious corners of the internet, answered in full and painfully researched detail. From the gaping maw of space to the most intimate crevices of human physiology, we will embark on this quest together.

I am damn determined to prove to you there are (no) stupid questions . . . in science.

BIOLOGY

"EVERY TIME I HEAR A POLITICIAN MENTION THE WORD 'STIMULUS,' MY MIND FLASHES BACK TO HIGH SCHOOL BIOLOGY CLASS, WHEN I TOUCHED BATTERY WIRES TO A DEAD FROG TO MAKE IT TWITCH."
—ROBERT T. KIYOSAKI, AUTHOR OF *RICH DAD POOR DAD*

HOW DO BACTERIA RESIST ANTIBIOTICS?

NO THANKS, I'M GOOD.

Starting this book off with one of the gravest concerns among public health officials? I would expect nothing less from this motley legion of curious minds.

Let's cover a few of the basics.

There are multitudes of molecular means that we can employ to take out a battalion of bacteria—different antibiotics utilize different mechanisms. For instance, we can destroy their protective cell wall (penicillins and cephalosporins), we can metaphorically castrate them and quell their ability to reproduce (fluoroquinolones, metronidazole), or we can knock out their capacity to manufacture certain molecules that may be essential to their survival (trimethoprim). In any case, whether we obliterate individual cells or their ability to make more of themselves, we use antibiotics to effectively terminate bacterial production, allowing your immune system to clear out the stragglers and make you feel all better.

But as simple as these organisms are, bacteria have become quite good at evading our molecular assassination attempts. Some of these bacteria may alter their own surface proteins so that the antibiotic compounds can no longer dock onto them, rendering them ineffective. Or they may even develop molecular pumps that literally shuttle the antibiotic compounds out of the cells as soon as they enter (which I've always found to be kind of hilarious—it's like a bacterium saying, "Sorry, can you send this back to the chef? I didn't actually order this.").

Now, you're probably asking yourself how the bacteria know how to do these things—do they hold microscopic OSHA meetings and determine that antibiotics don't provide them with a safe workplace? No. It honestly comes down to a game of genetic chance and a roll of the evolutionary dice.

Contrary to the enduring colloquialism, evolution is *not* "survival of the fittest." In actuality, evolution is more like "survival of the organism that can make the most babies." The word *fittest* is kind of a misnomer. Because bacteria divide so rapidly, there are plenty of opportunities for random genetic mutations to occur. These mutations either harm the bacteria or, quite often, are completely inert. But every great once in a while, a mutation will come along that confers an extra advantage by sheer dumb luck. In the context of antibiotic resistance, this may present as a physical change to the bacteria that disallows the antibiotic from doing its intended job. These bacteria can survive longer and thus reproduce much more effectively. As a result, they pass those same mutated-but-super-effective genes down to their progeny, forming a veritable army of resistant cells.

Even weirder is the fact that some bacteria can share trade secrets with other bacteria. So rather than conferring resistance only to its own offspring, a bacterium can also share with its neighbors! This is called "horizontal gene transfer" and is the sharing of genetic information *between* bacterial cells. The way this happens is either by direct injection of genetic material from one cell to another (called "conjugation"), a virus that picks up genetic information from one cell and transfers it to the next cell (called "transduction"), or a bacterial cell randomly chancing upon a chunk of genetic information that may be floating around (called "transformation").

Regardless of *how* these bacteria become mutated, antibiotic resistance is actually a pretty scary thing. I would be remiss in my duties as a scientist—and considering that one of my graduate degrees is in public health and epidemiology—if I didn't end on a PSA.

[Taps microphone for incoming PSA]

When humans expose bacteria to antibiotics unnecessarily, we unknowingly give them the cheat code to our defense systems. Today, antibiotics are overutilized; more often than not, a case of the sniffles is caused by a virus, the infection of which will rectify on its own in a few days. So, when people take antibiotics during these infections, not only are they *not* doing anything to help the course of their ailment (antibiotics do not work against viruses), but they're also giving bacteria one more chance to find resistance. Currently, antibiotic resistance is on the rise, and if you come into the hospital with a resistant strain that thwarts everything that can be given? There's really nothing that medical professionals can do for you, which can quite easily become a lethal situation. More and more of these bacterial strains are cropping up every year. So if you've got a case of the drippy nose, consider taking vitamin C and chicken noodle soup in lieu of penicillin . . . probably tastes a bit better, anyway.

HOW DOES DNA WORK?

Imagine the grandest library, dripping with French baroque. Its cavernous halls contain miles of the most precious human knowledge, organized into incalculable rows of tidy shelving. Due to the importance of the information contained within these books, very few people have access to the library, and clearance protocols are exceptionally stringent. No books are allowed to leave this place, only delicate photocopies of selected pages with explicit prior permission.

What you have just imagined is analogous to the nucleus of nearly every cell in the human body.

The nucleus is a small hollow pocket inside of the cell, specifically designed to safely house your entire genome. Every nucleus, inside of every cell, contains a copy of all your DNA—regardless of cell type and regardless of which parts of the DNA are being used by the cell. Each strand of DNA is a fraction of the width of a human hair and approximately six feet in length when fully stretched out.

Typically, the length of DNA is wound tightly around biological spools called "histones" to maximize space efficiency. But if you were to link every strand of DNA in your body end to end, from your trillions of cells, the total length would be approximately sixty-seven billion miles long.

. . . I know. Stunned me the first time I heard it, too. It's a staggering amount of information. So how does this little library of immense human information work?

The genome is a sacred library—it is a repository of how-to manuals, which contain all the schematics and instructions for building every protein in the human body. The genome provides the cell with blueprints for how to build little proteins, big proteins, proteins that can help perform chemistry

BIOLOGY 5

(enzymes), structural proteins, and proteins that link up with other proteins to form complex molecules that can carry oxygen through the blood or facilitate the electrical current that courses through neurons or, hell, even proteins that help copy the entire genome itself. In essence, the genome contains the blueprints for You.

Just like a sacred library, the information contained within your genome is protected and neatly categorized. Say I wanted to locate information on how to build a transport protein. Well, I would unfurl the DNA molecule to expose the section that codes for transport proteins (find the correct bookshelf), find the gene for the specific protein I need to construct (select a book), scan the gene for the information applicable to the specific protein I'm building (flip to the appropriate chapter), find the segmented genetic units associated with the protein build-out (locate the step-by-step instructions), and begin the process to create mRNA (photocopy the step-by-step instructions).

Information that leaves the nucleus may only be a *copy* of a section of DNA—never the DNA itself, which is important for protecting this all-important biological archive. This is the purpose of the mRNA photocopy—it acts to provide the instructions contained within DNA to the construction crew of the cell, which subsequently utilizes these instructions to build the associated protein outside of the nucleus. These proteins are assembled based on the instructions, piece by piece, until completion.

The process of reading DNA and the associated protein build-out is a bit like building IKEA furniture . . . if the instructions could occasionally be tens of thousands of steps long. And if incorrectly reading a step or two resulted in catastrophic failure.

So, yeah, I take that back—it's *exactly* like assembling IKEA furniture.

DOES CHICKEN NOODLE SOUP ACTUALLY HELP WITH BEING SICK?

The origins of chicken noodle soup can be traced back as far as the twelfth century with roots in multiple cultural cuisines from around the world. While I can't be certain, I'd also be willing to bet that mothers have been prescribing it to sick kids for just as long.

In many cases, cultural proverbs like "Chicken soup will make you feel better" hold little or no bearing in real-world application; they've become more of an expected verbal courtesy for the afflicted. However, you'd be surprised to learn that this salty broth may be implicated in immune support after all.

Several small studies have examined the impact of chicken noodle soup on the mitigation of upper-respiratory symptoms related to the common cold. Many of these studies have heralded some curious findings. One of my favorite studies was published by a research team at Nebraska Medical Center.[1]

This team evaluated the effect of chicken noodle soup broth on neutrophil chemotaxis (or in other words, whether the soup changed how neutrophils migrate). Neutrophils are a type of white blood cell—frontline soldiers that play an integral role in early immune system response. They arrive at the scene of infection, driven by distress signals from other cells, and once on-site, they launch a tiny war against invading microbes. But the chemicals released during this cellular battle cause inflammation. In the case of a viral common cold, the neutrophils begin to flood the lining of the respiratory tract, and mucous production ensues. So this team of researchers from Nebraska wanted to know if exposure to chicken noodle soup somehow helped reduce this neutrophil-guided inflammation. Their lab findings demonstrated that the soup did, indeed, significantly reduce the ability of neutrophils to find their target area. During the throes of the common cold, this might potentially contribute to a temporary reduction in neutrophil-related sniffles, snot, and coughing.

The best part about this experiment? The research team, within the methodology section of their manuscript, saw fit to include the detailed recipe they used for their lab soup. This was comprised of the ingredient list (which specified, among other things, three onions, several parsnips, and a six-pound stewing hen), cooking instructions, and "salt and pepper to taste." The team's technical name for it was "Grandma's soup," listed next to the lab protocol for neutrophil chemotaxis evaluation. And so Grandma's soup lives on into perpetuity within the hallowed halls of peer-reviewed scientific literature.

WHEN CATS PURR, WHAT IS ACTUALLY MAKING THE NOISE?

If you were to ask me, personally, why I think cats purr, I would say that it's some sort of low-frequency waveform produced from the animal's direct contact with a parallel, demonic dimension.

As you may aptly guess, dear reader, I am *not* a cat person.

If you were to ask feline experts why cats purr? You'd probably get multiple different answers, such as an expression of contentment, a social signaling attempt, or a mode of self-comfort. The truth is experts have no unified understanding of why cats purr.

Regardless of the rationale for the rumbling, the physiology that drives the purring mechanism is actually an intricate coordination of muscular contraction. Signals originating from the central nervous system are shuttled to muscle tissues located in two primary areas: the larynx (located in the throat and alternatively known as the "voice box") and the diaphragm (located at the base of the chest cavity and used to aid in the expansion and contraction of the lungs). The signals elicit rapid muscular contractions from both implicated muscle groups. These muscular contractions flutter at a speed of approximately 25 to 150 oscillations per second.

As the cat breathes, air moves past the vibrating structures in its throat and subsequently becomes disturbed at the same frequency. This is why you can not only feel the purring mechanism at work, but you can also *hear* a vibrating noise as well. This air disturbance is audible during both the inhalation and exhalation periods of the cat's breath cycle, which also gives the purring mechanism its distinctive continuity.

All that being said, I'm still going with the theory that cats are probably evil interdimensional travelers using their purring to lull us into a false sense of comfort.

HOW DO VACCINES WORK?

To understand how vaccines work, you need to understand some of the underpinnings of how your immune system functions.

But as a scientist, charged with the responsibility of combating widespread misinformation, I'd like to start with this:

First, there is *no* evidence to indicate that vaccines cause autism. The two studies that are typically referenced are both terribly flawed—one of the principal investigators actually lost his medical license and has since been academically discredited because all the conclusions in his paper, according to conclusions drawn by reviewers, were "utterly false." But thanks to the magic of the internet, his falsified works are still circulated. His name is Andrew Wakefield—look him up.

Second, mRNA vaccines do *not* alter your DNA. In fact, mRNA never enters the nucleus—it is biochemically banned from doing so. All cellular use of mRNA takes place *outside* of the nucleus, which is completely separate from the DNA. So it can't change your genome because . . . well . . . it doesn't even come into contact with it!

Third, no one is trying to secretly inject you with a transponder that would also record your conversations . . . chances are good that your conversations are quite boring anyway.

Now, back to the topic at hand. Essentially, your immune system operates via two separate squads: the nondiscriminatory *innate* immune system, which responds immediately to a threat, and the more refined precision assassins of the *adaptive* immune system, which establish a long-term recognition of foreign invaders. Together, the innate and adaptive systems collude to ensure you're covered, from your first exposure to a microscopic invader and into perpetuity.

When an unrecognized microbe enters your body, the innate immune system is the first to act. Its processes include a one-size-fits-all approach to dealing with the foreign invader: immune cells stumble upon a devious microbe, chemical SOS signals are sent out, and the cavalry arrives to help beat back the infection. During these initial stages, you likely have a drippy nose, or sore throat, or cough, or fever . . . you get the idea: you feel like hell for a while. As the innate immune system handles business, members of the adaptive immune system begin their own process of research and development: some of the cellular members of this team will engulf the invaders, chew them up into tiny bits, and give these tiny bits to other cells to study. This process is used to help the body establish a memory of what these microbes look like. In the long run, the adaptive part of the immune system is more efficient, more specific, and more rapid to respond, and it helps prevent you from becoming reinfected by the same microbes in the future. The problem is that the adaptive immune system takes about two weeks to finish its research and launch its battle campaign. In the interim, the innate immune system combats these infectious agents, keeping you from succumbing to them.

But, establishing that gold-standard, highly efficacious, adaptive immunity means that you have to become infected by the microbe in the first place, right?

. . . Not if you already know what these cooties look like.

This is what vaccines do: they provide your immune system with a description of the microbes. Subsequently, this allows your adaptive immune crew to study up on the assailant *without* you ever coming into contact with the real deal. It's like skipping the first part with the sniffles and sneezes and weeks of that weird lingering cough—you just fast-forward to the good part where your body manufactures WANTED posters for specific microbes, and biological bounty hunters start scouring the land for them. So when you *do* cross paths with that high-profile virus? Your body sees it, recognizes it immediately, and subdues it. This mitigates your symptoms, helping you to feel better much sooner (or in many cases, to feel nothing at all), and significantly decreases your likelihood of infecting other people.

When I was younger, I used to imagine that the cells of the immune system went to war with viruses and bacteria. But like, actual war. From the medieval period. With tiny horses and broadswords the size of cocktail picks. I would pretend that taking vitamins or medicine would provide more advanced armament to the immune cells so that they would ultimately ride into battle with bazookas and lasers. Coincidentally, at the time of writing this entry, I'm well into my fourth day of a nasty head cold. I am thirty-six years old and still imagine that same microscopic battle scene.

WHAT IS THE FUNCTION OF NIPPLES ON MEN?

Short answer: there is no biological necessity for man-nipples. Long answer: nipples on men teach us a valuable lesson in human embryology.

It might come as a bit of a surprise to you that the biological sex of human beings is assigned a bit late in the developmental game. When fertilization of an egg occurs, the zygote (or brand-new growing ball of amorphous cells, which will eventually form the fetus) comes equipped with all of the machinery to become male *or* female. Thus, the rapidly expanding human exists, for a short time, as a bit of both simultaneously. After a brief period of important developmental steps, the genetic, dimorphic switch is flipped. This will turn off characteristics of one sex (effectively silencing their expression) and turn on those of the other. However, the development of physical traits associated with biological sex occurs only *after* several morphological features of the body have already formed, including the rudimentary limbs, toes, fingers, and—you guessed it—the biological beginnings of nipples!

Interestingly, despite their lack of biological utility, male nipples are more than just skin-deep. In fact, men also maintain a network of mammary ducts and glands, which—had they been born female—would be fully operational for the nourishment of offspring. On occasion, due to certain pathological conditions (the origins of which can be benign or physiologically nefarious), male nipples may excrete fluid of various compositions. This leaky phenomenon can be clear, yellow, or even milky (a condition known as "galactorrhea").

While discharge and galactorrhea may be an indication of deficient testosterone levels or male breast cancer, it has also been associated with the use of performance-enhancing drugs in male athletes.

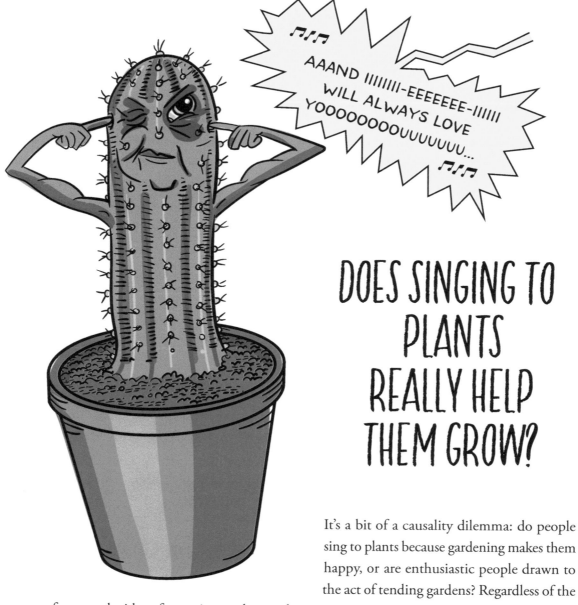

DOES SINGING TO PLANTS REALLY HELP THEM GROW?

It's a bit of a causality dilemma: do people sing to plants because gardening makes them happy, or are enthusiastic people drawn to the act of tending gardens? Regardless of the sequence of events, the idea of crooning to chrysanthemums is not a new one. In fact, Charles Darwin even suggested the possibility of plants exhibiting a growth response when exposed to mechanical stimuli—these stimuli could, theoretically, include vibrations from nearby sound sources.[2]

Despite the relatively wide acceptance of the singing-benefits-plants notion, the evidence is pretty paltry. Some studies indicate no relative benefit; other studies indicate that singing positively impacts growth.[3] Unfortunately, many of the experiments with positive findings seem to have limited sample sizes or quirky study designs—not especially reliable research stuff. So at this point, it's safe to say results are inconclusive at best.

But before you completely abandon those perfectly pitched Journey vocals (wow, am I dating myself with this reference?), allow me to offer you a bit of hopeful science. When you exhale, approximately 4 percent of your gaseous expulsion is carbon dioxide. This may not sound like an impressive amount, but it happens to be a concentration about one hundred times greater than the air you previously inhaled. Plants utilize carbon dioxide to manufacture glucose via little energy-producing factories inside of their leaves. So, theoretically, if you stood close enough to a potted plant and sang long enough and loud enough in a small area that wasn't well ventilated? Your exhaled carbon dioxide would probably contribute to a slight increase in plant energy production . . . assuming you also remember to water it. I always seem to forget that part.

HOW DOES A VIRUS MUTATE?

Viruses can replicate incredibly quickly. But they're also reckless in the handling of their own genomes, and they have—what I would call—a very clunky replication strategy. This means that most viruses can explode in number once seeded within a host, but they don't really have quality-control measures in place for ensuring that subsequent copies of themselves remain biologically intact—keep this in mind as we proceed.

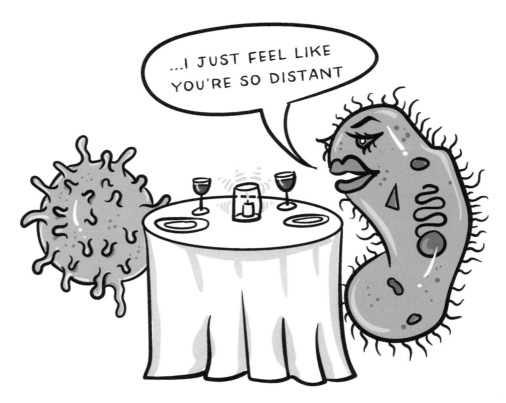

Viral evolution is what allows viruses to become more infectious, more virulent, and more resistant to vaccines. This evolution is driven by changes to the microbe's genetic material, either acquired through recombining its genome with the genome of other viruses or via good ol'-fashioned, haphazard mutation. We will focus on the latter.

When viruses replicate—much like a cell—their genome needs to be copied. One of these genomic copies is allocated to every subsequent bouncing-baby virus. During the copying process, random errors may occur by simple hiccups as the genome is scanned. But unlike human cells, viruses typically do *not* have the equipment to review these copies for errors. As a result, the errors go uncorrected and become hard-coded into the genetic how-to manuals for the build-out of new viruses. Sometimes, this can drastically alter the way the genome is read, and incorrect manufacturing of viral progeny ensues. This is what we call the "phenotypic mutation," or the big physical result of the gene change.

Unchecked errors can lead to an abundance of random mutations in the viral genome, which accumulate over the course of generations. Most often, the mutations don't impart much change. However, every once in a while, these mutations can confer an advantage to the virus.

You see, sometimes the genomic misspellings aren't complete gibberish—by pure dumb luck, they may slightly alter something about the virus that makes it better at infecting its target host. Now, this could be a small change to the shape of a protein on the surface of the virus that helps it more easily dock onto a host cell. Or maybe the mutation changes something about the way the virus looks, which gives it a little incognito flair around the immune-system sentries.

Ultimately, if the mutation helps the virus to be more of an infectious scourge, then it has a greater opportunity to reproduce and continue its lineage. But again, this is all a product of a spin of the genetics roulette wheel, which *all* organisms and microbes are subject to. Sometimes, viruses can get lucky and acquire the ability to make you sicker; I was unlucky and acquired strange-looking toes.

WHAT HAPPENS TO SPERM AFTER FERTILIZATION?

Sperm cells may be the ultimate authority on valiant, self-sacrificial acts. To complete the mission of passing on precious genetic code, sperm cells are created simply to be destroyed—their mad dash to the finish, punctuated by an abrupt end. The winner of this biological race is decapitated; the remaining millions are forced to swim about aimlessly, killing time as they wait for their own demise.

While it may sound like a plotline extracted from a sad Russian novel, it is—in fact—incredibly biologically accurate. And pretty cool.

It goes something like this: After ejaculation, approximately twenty million to three hundred million sperm cells are released. Like a (very) large pack of bloodhounds, these cells navigate their way to the egg by following a trail of chemicals, effectively "sniffing out" the location. Once on the scene, they swarm the egg and bang against its outer layer, dumping digestive juices onto the surface. These juices eat away at the egg's feathery glycoprotein cover, boring a hole through it in an effort to create a passageway to the inner plasma membrane of the egg. The first sperm cell to successfully make it through a hole and fuse itself with this plasma membrane wins, the rest of the team disbands, and the race is over. Ultimately, there is only one winner out of the millions of initial contenders (ideally).

What happens to this winning sperm? Well, it doesn't get much time to bask in the glory, I can tell you that much. Shortly after the fusion process takes place, the sperm's payload—the genetic material inside—is dumped into the egg for genetic combination and eventual embryo formation. At this time, the sperm tail is broken off, and all the parts of the sperm that are *not* implicated in fertilization are broken down. As for the millions of losing counterparts, they can no longer enter the egg and so arbitrarily cruise about the female reproductive tract until they run out of gas and die.

Fun fact: sperm cells can actually end up in the abdominal cavity! There is a small gap between a woman's ovaries and the opening to her fallopian tubes (which are surrounded by little finger-looking projections called "fimbriae"). While sperm cells aim to meet the egg within the fallopian tubes, some sperm cells may overshoot this target and leave the fallopian tubes altogether, only to swim around in the uncharted waters of the abdomen.

So, to circle back to the primary question . . . there are really two different endings for our poor little sperm cells: they can win and become degraded, or they can lose and become degraded.

An existential crisis on the tiniest of scales.

WHY ARE MEN TALLER THAN WOMEN— WHAT PURPOSE DOES IT SERVE?

One of the coolest things about being a scientist is that my field evolves. As do all fields of science. You see, as we understand more about the underpinnings of how the world works, we update the Grand Repository of Scientific Knowledge about Stuff to reflect this elevated level of understanding.

For the record: in the context of life and existence, et al., we still understand very, *very* little. Anyway . . .

The differences in the physical appearance of organisms based on sex are collectively called "sexual dimorphism." This term is the big sticker we slap on certain species to tell us that lion manes are seen on males or that female oak toads are identified by the dark spots on their bellies. And, yes, human beings are included on the sexually dimorphic guest list.

Originally, it was postulated—and widely accepted—that the reason for the *Homo sapiens* male

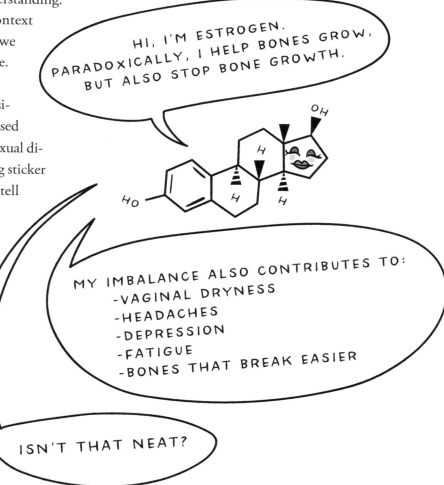

to exhibit a larger stature than the *Homo sapiens* female was driven by the behavior of the latter. Simply put: we thought that ladies sought out bigger and bigger boys for what was assumed to be the protection of family and higher competency in hunting or fighting. This would have theoretically led to an increased reproductive advantage and—*poof!*—after thousands of generations, men would have become generally larger. However, quite recently, this idea has been challenged by our better grasp of human biochemistry.

The newer, more direct rationale for the size difference? It may come down to the different hormone cocktails coursing through our respective bloodstreams.

As it happens, a surge of estrogen (the female sex hormone) is quite good at rapidly elongating bones during early puberty. However, as puberty draws to a close, estrogen is also implicated in sealing up growth plates (the areas at the ends of long bones that house bony stem cells and where the growth is derived from). In fact, this short-lived, estrogenic sprouting of bones is precisely what we witness during female puberty, directly before menstruation cycling begins. It's also why middle school–aged girls are often taller than their male counterparts. But after female growth plates quickly seal themselves shut and growth ceases, male maturation steadily drives their bone growth (and height) upward.

I myself was dealt a crushing endocrine-based blow to my ego. In seventh and eighth grades, I was always positioned near the back of the class photos—the only acceptable place for the tall (superior) kids. On my first day of ninth grade, as a brand-new high school guppy, I was devastated to find that I suddenly had to crane my neck upward to peer into the faces of several of my childhood, male friends. Talking about you, Brian—you six-foot-eight bastard.

IS NOT HAVING SEX HARMFUL?

If you were to poll the public for their personal opinions on the matter, I'm sure *some* people would instantaneously blurt out a resounding "YES!"

. . . It's me. I'm some people. But! Biases aside, let's take a crack at the actual science.

It is of incredible importance that we begin this analysis with a few facts: having sex is a personal choice, as is not having sex. The act itself should be consensual and safe. The appropriate frequency, or infrequency, of sex is unique to every individual on the planet.

Now, to the question at hand. The empirical evidence on this matter seems to indicate there are both benefits and drawbacks to a dry spell. Let's get to know these things a bit more . . . *intimately*.

(I'd like to formally apologize for that joke.)

The first benefit of abstinence (whether it be intentional or circumstantial) includes a decreased risk of infection. It may come as little surprise that getting down and dirty is, well, pretty dirty. If

you want to talk numbers, up to seven hundred known species of bacteria may populate the urinary tract, with an additional seven species of fungi that may tag along as well. The skin of the penis plays host to more than forty species of bacteria, while the vagina also harbors more than twenty distinct kinds of *Lactobacillus*. While most of these bacteria are an integral component of keeping your private bits healthy, the bumping together of genitalia *is* associated with a higher risk of developing a urinary tract infection and may also be associated with bacterial vaginosis. Turns out, mashing bacteria into areas where it doesn't belong may not work out so well.

While the data appear to be a bit sparse, some studies have also suggested that a lack of sex can help boost mental concentration.[4] The theory is that focusing on professional tasks, or even personal growth, is made easier when the all-powerful, evolutionarily hard-wired drive to reproduce is put on mute.

As for documented physiologic pitfalls, going sexless may be associated with a less robust immune system, higher anxiety levels, disrupted sleep patterns, and a possibility of increased risk of prostate cancer in men.

At the end of the day, there is a laundry list of give-and-take when evaluating a sexually active, or decidedly not sexually active, lifestyle. But no grave harm seems to be on the horizon if you find yourself in a scenario of celibacy. There's also the added benefit of being able to make balloon animals out of any unnecessary condoms you might have floating around.

ARE GMO FOODS BAD FOR YOU?

I get the impression that many people have misconceptions regarding what GMOs are and what they are not. Probably best to do some clarifying first.

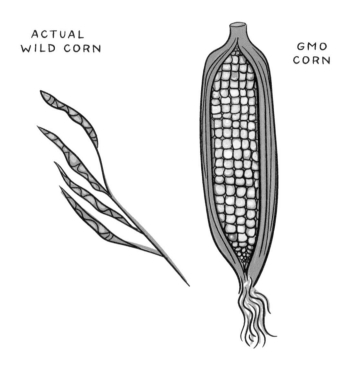

GMO is an initialism for "genetically modified organism." This term describes *any* organism that has had its genome intentionally manipulated; the group includes bacteria, fungi, roses, mice, fish, frogs, flatworms, pigs, and, of course, food crops. While this might sound like some scary, science-fiction situation—with faceless figures in lab coats brooding over steaming vats of Soylent Green—it is actually a very old practice that is partially responsible for the boom of human civilization.

Genetic modification has been ongoing since antiquity—we're talking historical dates in the BC era. As groups of humans became less nomadic, they established homesteads and began experimenting with crop growth. A great example of modified food from this part of history comes from corn.

If you've never seen what real corn looks like, you wouldn't recognize it. The succulent yellow cob we know today has been selectively crossbred over thousands of years. The original version? A small weedy stalk with a braided row of sad, fibrous pods on one end. But human beings engineered it to produce the glistening golden kernels that grace every backyard BBQ.

Modern GMOs aren't simply utilized to make bigger produce, though. An immense number of important modified crops have been developed to do wildly useful things. Some of them resist nuisance insects that would otherwise decrease food yields; other experimental crops can grow in the face of drought (something that is becoming increasingly important to us as global warming advances).

Now, unless you've lived completely off the grid for the last couple of decades, then you know GMO crops have received a bad rap. If you do some digging, you will find that allergic reactions, cellular toxicity, and possible cancerous behavior have been reported in some preclinical studies in which animal models were fed GMO foods.[5] These concerning findings are associated with the proteins produced from the modification itself. You see, the way we genetically modify food is to add a new gene to the genome of the crop of interest. A new protein is produced from the added genetic information, which alters something about the crop—this could be anything desirous, from beefing up its nutritional value to giving it the ability to resist herbicides. However, these good intentions alone are not enough to ensure hitch-free genomic editing.

Let's say you want to make rice more nutritious to help address problems with nutrient deficiencies in remote villages—a noble cause, no doubt. So you decide to insert a gene that produces a protein implicated in beta-carotene production. Soon you're producing superfood rice and solving malnutrition, ready to collect your accolades from public health officials worldwide, right? Not exactly. The manipulation of DNA isn't as easy as cutting and pasting. In fact, inserting a gene or two could have unanticipated consequences. Humans are pretty new to the whole genomic editing game, and we've been learning that gene insertion has the possibility of changing how other native genes may be expressed, or it may change how protein products from neighboring genes are built. While some of these changes may have no effect at all, accidentally altering the *wrong* things could lead to outcomes like flipping on an oncogene (a gene associated with cancerous activity) or shutting off a part of an important metabolic process. Not good.

This is why preclinical models and lab testing are so important—before clearing GMO foods for human consumption, we first need to know if we've done anything deleterious. Currently, there are regulatory bodies in place to fervently monitor the development processes and safety of these foods, including the USDA and FDA (United States), European Food Safety Authority (continental Europe), Office of the Gene Technology Regulator (Australia), and many more. Additionally, groups like the

American Medical Association have begun to help synthesize the data from GMO studies to advise on human health and safety as well.

All right, we've made it this far, so let's go ahead and answer the question: are GMO foods bad for you? Currently, we cannot responsibly make sweeping generalizations and say that *all* GMO foods are safe—there are simply far too many nuances for us to know all the risks associated with every crop combined with every different genetic modification. That being said, several GMO foods have already met appropriate safety standards and are currently in the market—foods you may have already eaten, like soybeans, sugar, beets, corn, and a few others.

You may be wondering, dear reader, would *I* eat a genetically modified apple? Sure, if it had been thoroughly vetted for safety. But, then again, I wouldn't consume, or drive, or wear, or touch, or drink anything that was associated with unknown health risks.

The reality is we live in a brave new world. It's no secret that feeding the 7.7 billion (and growing) people on this planet is going to be increasingly difficult. So, while genetic modification is experiencing some growing pains during its infancy, its refinement *does* have the potential to help our species exist in the face of a growing list of unprecedented challenges. Just like developing penicillin or chemotherapy or myoelectric prosthetic limbs, we must continue to hone the science until we can ensure its efficacy. It's too important not to.

And if we ultimately can't ensure that genetic modification is completely safe, I'd settle for it giving me some off-brand superhero capabilities.

ARE VIRUSES LIVING ORGANISMS?

Viruses: easily forgotten until the scourge of flu season rolls around. These little microbes can be highly infectious, and they're hell-bent on transforming your cells into their own personal virus-producing factories. So it may come as a bit of a shock when I tell you that viruses are, in fact, *not* living organisms.

[Cue: record skip]

I know, I know! You're sitting there, joyfully reading through this fun book you picked up, and now you're seriously questioning the credibility of your scientist-author. But I assure you that I can explain the rationale for this nonliving microbe (while also maintaining my trusted reputation as an authority in these matters).

In general, to be considered among the living, organisms must meet a certain set of criteria. This includes the following:

1. INTERNAL ORDER AND ORGANIZATION
2. THE ABILITY TO RESPOND TO OUTSIDE STIMULI
3. GROWTH AND DEVELOPMENT DURING THE LIFE CYCLE
4. THE HARNESSING OF METABOLIC PROCESSES
5. THE CAPACITY TO MAINTAIN HOMEOSTASIS
6. THE ABILITY TO REPRODUCE

This list will help guide us in our classification. So let's work our way through each item.

ITEM 1: Viruses possess no internal structure, no tiny organs, and no cellular compartments that are typically imperative in running a living biological system. Because living organisms host a symphony of complex biochemical reactions necessary to maintain life, they must have small compartments to effectively separate them. Viruses have squat.

ITEM 2: Viruses don't have sensory receptors—they cannot see, hear, smell, touch, or taste. They are simply a protein bubble with some genetic material stuffed inside (some of them come in fancy varieties that may have an enzyme or two as well). So they can't respond to stimuli from the external world.

ITEM 3: Viruses do not grow, they don't change forms, and they don't really mature. Again, we're talking about a simple protein bubble with some random chunks of genetic code contained within.

ITEM 4: Viruses cannot harness or utilize energy . . . for any processes. Living organisms require energy to fuel the myriad of mechanisms that sustain their existence. Viruses have absolutely no machinery to acquire or harness energy of any kind.

ITEM 5: "Homeostasis" is the consistent maintenance of certain biological parameters, including metrics like pH levels, temperature, fluid composition, etc. Viruses have no ability to regulate these—the viral body gets what it gets and is completely at the mercy of the environment around it.

ITEM 6: This is the biggie: viruses cannot reproduce on their own. At all. In fact, they are so inept when it comes to reproduction that they go to great lengths to hijack all of *your* cellular equipment in order to multiply. They are the quintessential intracellular parasite: they cannot propagate without utilizing what you've got.

So, if viruses aren't alive, what the hell are they? Turns out they're just a random assembly of barely functional molecules . . . similar to several of my exes.

DO MEN HAVE A BIOLOGICAL CLOCK?

The mention of a "biological clock" has become colloquial, and the term often refers to the notably strong, longing pangs a woman may feel about wanting to have children. However, when referenced in science and medicine, the biological clock is less like a call to action and more like a countdown to reproductive doomsday.

The biological clock is, more or less, another way to reference the projected remaining time left in an individual's viable fertility. For women, fertility typically peaks from ages twenty to thirtyish (depending upon which research studies you reference) and then begins to decline until age forty, with a sharp drop-off shortly thereafter. The medical community used to be under the impression that men were not subject to similar reproductive sanctions. But recent evidence seems to suggest otherwise.

Multiple studies have begun to peer into the implications that poor sperm quality may have on the developing fetus.[6] As men age, testosterone levels take a nosedive. This hormonal imbalance may contribute to a corresponding drop-off in sperm production. Additionally, advanced age is associated with a higher risk of accumulated damage to DNA—something which can adversely affect what sperm cells look like and how they function. Taken together, this means that the biological conditions of older men can often contribute to the production of sperm with crooked tails, misshapen heads, or genetic abnormalities. These changes have been associated with nonviable pregnancies and—while not well understood—even an increased risk of the child developing schizophrenia.

The biological rationale for potential age-dependent fertility caps in men is still being unpacked. So while we have some work to do to understand more about fertility changes in males, rest assured that they likely *do* have a biological clock.

My apologies to any young men out there who have asked their eager partners to wait to have children. My intent was not to throw you under the bus with this one . . . as far as you know.

WHY DOES EVERYONE SAY BEES DYING IS A PROBLEM?

The honeybee crisis (which is, surprisingly, *not* the name of a progressive folk band) began in 2006 with alarming mass reports of hive collapse around the United States. The cries of panicked beekeepers rang out all over the country in response to the somewhat instantaneous disappearance of their domestic colonies. It didn't take long for the international community to recognize their similar loss of bees—both domesticated and wild. As a result, multiple countries began to dedicate scientific resources to studying this disconcerting phenomenon.

The official term for the widespread loss of bees is now called "colony collapse disorder." During the first decade since it was recognized, approximately 30 percent of colonies have been reported lost every year—this shakes out to *double* what the expected losses should be due to normal, natural causes.

Teams of researchers have been working diligently to understand the causality behind the concerning spike in bee mortality. Unfortunately, there is no cut-and-dry rationale for the worldwide loss. A multitude of reasons have been put forth, and they all seem to play a devious role in the detriment. This complex disorder may be caused by parasites, loss of wildflowers, pesticides, and pollution. The common physiologic reason boils down to this: Honeybees have an elegant neural circuit that provides them with precision spatial

memory, sharp sensory mechanisms, and incredibly accurate internal GPS. External stresses are scrambling their tiny brains, which makes it impossible for them to locate food or find their way home after they're done collecting nectar—a death sentence for a bee.

As teensy as they are, bees shoulder an enormous amount of responsibility. Not only are honeybees adorable (seriously—they look like they're wearing tiny fuzzy vests), but critical pollinators, like bees, are also implicated in up to 80 percent of US-based crop production. Their activities not only help us sustain our crop yield, but they also support a multitude of other organisms associated with that food chain as well. As we lose more and more of these pollinators, their decline makes crop production more difficult, drives up the cost of farming, and drops the volume of available produce. This could lead not only to crop shortage but to economic ramifications from inflated food prices as well.

So, if you've got a fear of bees, maybe take a minute to reconsider your preconceived notions. The reality is that a *lack* of bees is far scarier.

ARE THERE ANY DISEASES THAT HAVE JUMPED FROM HUMANS TO ANIMALS?

As a brooding childhood scientist, I used to worry that my intermittent cases of sniffles would be passed to the family dogs. When I was sick, I used to barricade them out of my bedroom and away from my tissue-littered campsites on the couch for their own protection. In hindsight, the dogs probably thought I was being a complete jerk. In actuality, I was just trying not to communicate my seasonal illnesses to them.

I'll be the first to admit, at that age, I obviously had zero understanding of the interspecies transmission of the common cold. But interestingly enough, I wasn't completely off the mark. When animals pass diseases to humans (like some of the headlined swine or avian flus), it's called "zoonotic" transfer; when we pass diseases to animals, it's called "reverse zoonotic" transfer. Just figured you'd want to know the technical terminology in the event you feel like impressing your friends.

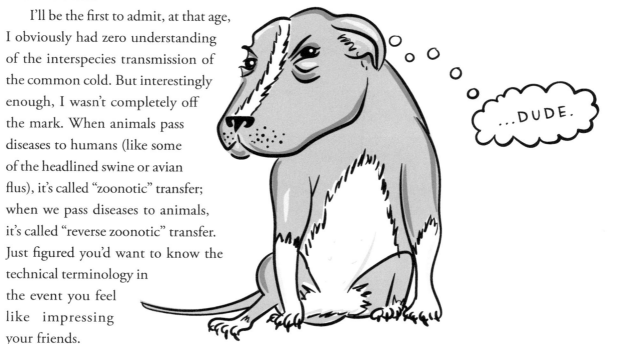

Reverse zoonosis occurs more frequently than you may realize. Here are some of the more bizarre cases:

- CAPTIVE ELEPHANTS THAT CONTRACTED TUBERCULOSIS IN THE HAWTHORN CIRCUS CORPORATION
- FERRETS THAT TESTED POSITIVE FOR COVID-19 IN A WUHAN, CHINA, SEAFOOD MARKET
- *HELICOBACTER PYLORI* BACTERIA (WHICH CAUSE STOMACH ULCERS IN HUMANS) THAT WERE FOUND IN A TINY AUSTRALIAN MARSUPIAL CALLED THE DUNNART
- MOSQUITOES GETTING MALARIA FROM HUMANS (AT WHICH POINT THEY FLY AROUND, BITE OTHER HUMANS, AND TRANSFER MALARIA BACK TO US—ONE BIG ZOONOTIC–REVERSE ZOONOTIC INFECTION CYCLE)

So that I don't scare you with information about cavalier microbes leaping between species: a complex sequence of biology needs to line up just right to make zoonotic and reverse zoonotic transfer happen. It's definitely not an easy sell, pathologically.

So that I *do* scare you: a team of researchers from the Global Virome Project estimated there may be more than eight hundred thousand viruses yet undiscovered that could make the jump from animals to humans.[7]

I'm still kissing my dogs on the face, though—worth it.

IS HONEY ACTUALLY BEE POOP?

You will be relieved, dear reader, to learn that honey is not bee poop. Instead, honey is closer to bee *puke*.

See? Don't you feel better?

Honeybees buzz around, from flower to flower, happily slurping up sweet nectar through their long, hollow tongues. After a long day of sipping, they fly back home to their central hive to deposit the goods. The nectar inside of each bee's little belly (which is actually called a "crop") gets churned and mixed with various enzymes that act like preservatives to keep the nectar fresh during storage.

Cool. So the honey is made inside of tiny bee stomachs, right? Wrong. It's way weirder.

The bee who collects the nectar (we'll call this the delivery bee) returns to the hive with a gut full o' sweetness. The delivery bee doesn't just proceed to the deposit box, though. The delivery bee instead pukes the enzymatic nectar into the mouth of a second bee. The second bee swallows and swishes this puke, only to regurgitate it into the eager mouth of a third bee. This chain of upchucking proceeds like this until the nectar is spit into an empty wax pocket of the honeycomb. The adjacent bees then beat their wings to fan the nectar, hastening the evaporation process and thickening the slurry, and you end up with the sticky golden goodness that you apply to your morning toast.

The lives of honeybees: short in duration, but complex and confusing in practice.

IF HUMANS COME FROM MONKEYS, WHY ARE THERE STILL MONKEYS AROUND?

Ah, yes—a common question from those who deny the plausibility of evolution. This query originates from a misinterpretation of our grand timeline of events, as well as a misunderstanding of the mechanism of evolution. So I am more than happy to help clarify. The explanation is actually quite simple:

Humans did *not* evolve from monkeys.

Humans also did *not* evolve from apes.

Humans share a common ancestor with monkeys, and with apes, from many millions of years in the past. So, rather than monkeys and apes being part of our immediate nuclear family, we're more like distant (quite distant) cousins . . . who see each other occasionally at weird family reunions . . . and we kind of just nod at each other in recognition.

It goes like this: Approximately twenty-five million years ago, there was an organism that predated monkeys, humans, and apes—this was the great-grandmother of us all and the species from which we all originated. As time barreled forward, different species began to branch off as they found their own specialized biological niches. You see, new species develop as animals adapt to best survive in their highly specific areas. Every time they change a little bit to meet the demands of survival—based upon their geographic location, food availability, weather

BIOLOGY 41

patterns, predator index, etc.—they become a little different from the animals they originated from. With enough change over time, these organisms become *so* different that they are classified as a completely new species.

Monkeys, apes, and humans represent deviating groups that found their own niches. Monkeys were the first to deviate, then orangutans, then gorillas, then chimpanzees, then humans—we didn't originate from each other at all; we evolved in *parallel* based upon our species' respective, individualized needs.

This means monkeys still exist because they developed differently *alongside* us (just as birds and fish and chunky raccoons have). We may have more in common with monkeys and apes than with, for instance, a sea turtle. However, this is only because we shared a common ancestor far more recently. Monkeys and gorillas and chimpanzees still exist because they are different creatures. Altogether.

Now, the follow-up question that scientists are typically asked is this: why didn't chimpanzees cognitively develop to invent the light bulb, or wear patent-leather shoes, or ponder the origins of the universe? The honest answer is that they didn't need to—they adapted to their environment so well that there was no evolutionary pressure to continue to change. They found their niche and thrived. And that is where they remain today.

There is also a strong possibility that they saw us working nine-to-five cubicle jobs and said, "Nah, that's okay—we're good."

WHAT IS THE PURPOSE OF A UVULA?

The hanging lump of flesh that dangles into the opening of your throat is formally called the "palatine uvula." Beyond being an odd-looking addendum to your soft palate, it also contains muscle fibers and can rapidly produce saliva. While the precise rationale for its niche existence is still being explored, there are several commonly accepted theories to explain this weird little throat ornament.

The anatomical location of the uvula puts it at the back of your soft palate. As you swallow, the uvula and soft palate move upward in tandem to seal off your nasopharynx (the space behind your nose). This keeps your half-chewed hamburger bits from traveling upward and into your sinuses during the swallowing phase. This movement of the uvula has also been associated with assisting the former behavior of our ancestors when they would bend down to drink out of murky streams—the uvula acting to prevent water from sloshing back out of the parched human mouth and nose.

In addition to its mechanical role in swallowing, investigators have also studied its importance as a less commonly known immune system sentry. In fact, the uvula harbors its own little army of pathogen-fighting cells, likely used to recognize and sound immunologic alarms in response to ingested microbes.

Anatomists have also formulated the idea that the uvula has stuck around, evolutionarily speaking, to facilitate speech patterns. Human beings are incredibly vocal, and the noises we make are highly varied. Many of our languages (including dialects in French, German, and Arabic) include pronunciation that evokes uvular rattling (I actually just made that term up, on the fly), wherein consonant noises are articulated in the back of the throat. So that guttural, gargling utterance you couldn't quite master in your Russian 101 course? You can blame your uvula for that.

WHAT IS A TARDIGRADE?

All right, dear reader, I'd like you to do something for me. Take fifteen seconds and think of an example (or two) of what you would consider the world's toughest animal.

Excellent. Now, let us peruse your possible candidates. Some of you probably chose a big cat—a lion or a tiger perhaps. I'm willing to bet that some of you thought of a grizzly bear. For my readers located in coastal areas, you may have even considered the ruthless orca. While these are all incredible guesses, they are all *wrong*. By a long shot. In fact, the world's toughest animal cannot be readily seen with the naked eye.

Tardigrades are exceptionally small superheroes and are better known by the household name "water bear." They are their own phylum of microanimal, approximately one millimeter in length, with a pudgy body, four pairs of chunky legs, and a mouth that looks a bit like the business end of a trumpet . . . if a trumpet had sharpened spears poking out of it.

Tardigrades are found in nearly every location on the surface of the earth—we're talking snowy mountaintops, crushing ocean depths, lush rain forests, glistening sand dunes, and an array of freshwater lakes. They've even taken up residence in the ice of Antarctica, as well as volcanic sea vents. But far more astounding than their wide variety of real-estate choices are the conditions in which they can survive.

Ready for this?

BIOLOGY 45

Tardigrades can survive

- EXTREME TEMPERATURES RANGING FROM -328°F (-200°C) TO 300°F (151°C),
- BEING IMMERSED IN BOILING ALCOHOL,
- EXPOSURE TO IONIZING RADIATION MANY MAGNITUDES HIGHER THAN WHAT HUMANS CAN SURVIVE,
- BEING COMPLETELY FROZEN FOR DECADES,
- THE VACUUM OF SPACE, AND
- BEING SHOT OUT OF A GUN AND EXPERIENCING A SUBSEQUENT IMPACT VELOCITY OF ABOUT 3,000 FEET PER SECOND (900 METERS PER SECOND).

The grandest feat of all? This miniature beefcake has had the biological robustness to survive *all five* mass extinction events on Earth. Can't say the same for the front-runners you chose at the beginning of this entry, huh?

Due to its ability to withstand such extreme conditions, the tardigrade is also an interesting player in a theory called "panspermia." Panspermia is the scientific hypothesis that life was seeded on this planet from elsewhere in the universe, carried here by an asteroid or other space debris. Because of this little creature's resistance to the vacuum of space, extreme heat, and massive impact pressures, many scientists believe they would also likely survive a fiery trip to Earth, riding atop a clump of space rock. Once they crash-landed, some of them might have been viable enough to kick-start the humble beginnings of life.

I hope that this entry has inspired in you a measure of awe for these possibly-alien-in-origin microcritters. I love tardigrades. In fact, my left flank bears a tattoo of a rotund pink tardigrade, enveloped in a prominent banner that reads, "Tardigrade Tough."

Look, I only said I was a scientist. I never said I was cool.

IS IT POSSIBLE TO CLONE A WOOLLY MAMMOTH?

Let's just come out of the gate swingin'. Yes, we do have the capability to do this. But before we decide to re-create the worst parts of Jurassic Park, we should chat about it.

There are several techniques used to clone organisms, and these techniques are currently used in laboratories worldwide. Research teams use cloning as a means to copy genetic sequences they find to be helpful. Whether or not a genomic sequence is useful to a scientist really depends on the nature of their research. Scientists involved in agricultural studies may be drawn to engineering livestock with leaner bodies, while scientists who study a particular disease may be interested in mice that exhibit disease-specific genetic defects. Regardless of which desirable traits the DNA sequence confers, cloning allows us to make exact copies of a genome to support the continuum of research in a given field.

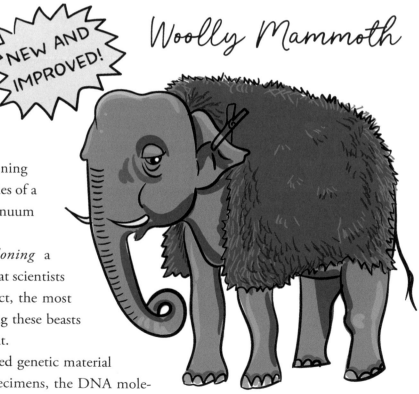

However, actually *cloning* a woolly mammoth is not what scientists are attempting to do. In fact, the most realistic method for bringing these beasts back is far clunkier than that.

While we *have* uncovered genetic material from woolly mammoth specimens, the DNA molecules themselves are usually pretty beaten up. Having spent tens of thousands (if not hundreds of thousands) of years suspended in permafrost, the genetic material we dig up is exceptionally fragile. These damaged DNA strands are not good candidates for cloning, but their sequences can still be read, and we now know the woolly recipe. So rather than cloning a mammoth with its own broken DNA, scientists are trying to *make* a mammoth, using that same recipe, by altering its closest living cousin: the Asian elephant.

It's a bit like a biological version of the old show *Pimp My Ride*, where the rapper Xzibit would show up at people's homes and customize their dilapidated cars. However, rather than sticking a set of spinning rims onto an elephant, scientists are attempting to give it more fat mass, smaller ears, and—of course—a lush woolly coat. When this proposed organism is created, and if our genetic engineering is successful, it will look exactly like a mammoth. But it won't be a mammoth—it will be a genetically tricked-out modern elephant.

The scientific community is split on the ethics of this experimentation. The group in support of the idea argues that mammoths could play an important role in slowing the rate of global warming, due to the important ecological niche they held in Arctic regions. The group in opposition thinks that the environmental rationale is an excuse to revive an extinct animal. Many scientists also argue that millions of dollars in this field of research would be better used preserving modern species that we are close to losing, rather than trying to resurrect those that have been gone for thousands of years.

Now, I think any scientist might admit (if they were out of earshot of their ethics boards) that the idea of resurrecting an extinct animal is pretty exciting—an incredible feat and a testament to our scientific advancement, to be sure. Especially a woolly mammoth, the strong and stoic mascot of the Ice Age. But just because we *can*, does this mean that we should?

We are infants of the cosmos, blinded by our momentary excitement of self-proclaimed grandeur. How much we have yet to know, and how small our mark among the stars.

CHEMISTRY

"IF I WERE TO TAKE AN UNDERGRADUATE CHEMISTRY EXAM,
I WOULD PROBABLY FAIL."
—VENKATRAMAN RAMAKRISHNAN,
RECIPIENT OF THE 2009 NOBEL PRIZE IN CHEMISTRY

WHY DO HOT PEPPERS BURN MY MOUTH EVEN WHEN THEY'RE COLD?

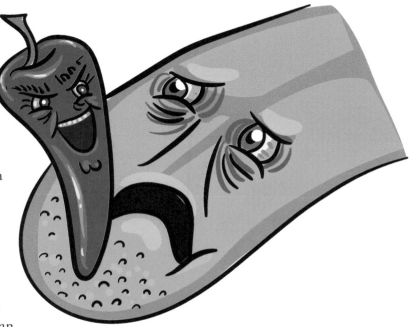

In doomed demonstrations of gustatory strength, we've all made the grave error of eating raw peppers in front of a cheering (or wincing) group of friends. In my case, I was alone and had recently harvested a handful of small Thai peppers from a potted bush on my balcony. Each pepper was, at maximum, one half inch in length. Because they were so small, I didn't see any obvious harm in sampling one or two of them. After all, I grew up in Mexican neighborhoods in Southern California—I live for spice.

It took approximately sixty seconds for my mouth to feel like the inside of a blast furnace. Within twenty minutes, I was flopping about on the cold tiles of my bathroom floor, pathetically clutching my stomach.

The throbbing burn from spicy food is not a product of temperature, despite the uncanny sensation of hellfire. It is, in fact, defensive chemistry deployed by plants to deter hungry passersby from eating them. The chemical culprit is a molecule called "capsaicin," which is produced in multiple areas of the pepper itself. Upon contact with this oily substance through ingestion or touch, capsaicin will interact with sensory neurons in the skin that are responsible for communicating signals of heat and pain to the brain. The molecule docks onto receptors found on the surface of the neuron itself (specifically, the receptor TRPV1), which triggers a pulse of communication from the pepper contact site to the brain. This signal says, "Ouch, ouch, ouch, ouch, OUCH! Something's hot! It's burning!"

Sure, inadvertently coating the inside of your mouth with ouchy burn molecules can be unpleasant. But imagine the point of view of pepper plants. Through the course of millions of years of painstaking evolution, they finally develop a chemical defense system to cause pain to creatures that risk destroying their seeds. Only to find one particularly cavalier, hairless ape that uses these chemicals as a complement to tortilla chips.

HOW DO BAKING POWDER AND BAKING SODA WORK?

Certain acid–base chemical reactions can yield a substantial plume of carbon dioxide gas. You might intimately understand this if you have ever been lucky enough to attend a middle school science fair. Every one of these events hosts at least one erupting volcano model, the tabletop mechanism of which is a simple reaction between vinegar and baking soda.

Baking soda (otherwise known as sodium bicarbonate) is a base. It is used in recipes that include acidic ingredients (in the world of baking, this might be buttermilk or citric acid of some kind). When you mix the dry ingredients with the wet ingredients, the acid and the base come into contact with one another, and a chemical reaction begins.

When a carbonate compound (like baking soda—remember, it's called sodium bicarbonate) is ex-

posed to acid, there is an exchange of electrons as the acid and base attempt to neutralize each other. One of the products that come out of this tiny chemical battle is carbon dioxide gas. I could show you exactly why this happens by writing out the chemical equation, but that would be super boring for most of you. So just take me at my scientific word on this and spare yourself the snooze fest.

When you're preparing your grandmother's treasured chocolate-chip cookie recipe, you are driving an acid–base reaction within the dough. Tiny bubbles of the resulting carbon dioxide gas begin to diffuse through the delicious mixture, and the cookies fluff up during the baking process. If you've ever run out of baking soda and tried to make cookies without it, you already know that you end up with super flat, super hard sugar pucks.

Baking powder applies the same principles with a slightly different delivery mechanism—it already contains both an acid and a base. Once any liquid ingredient from the recipe is mixed with the baking powder, the acid and base interact under similar chemical processes observed with baking soda. Again, carbon dioxide evolves, and your delectable dessert becomes delightfully puffy. Because baking powder already contains both the acid and base compounds, it's typically used in recipes that do not already call for an acidic ingredient.

All this talk of pH neutralization and the diffusion of carbon dioxide gas makes me want cookies, you know what I mean?

HOW DOES GLOW-IN-THE-DARK PAINT OR STICKERS WORK?

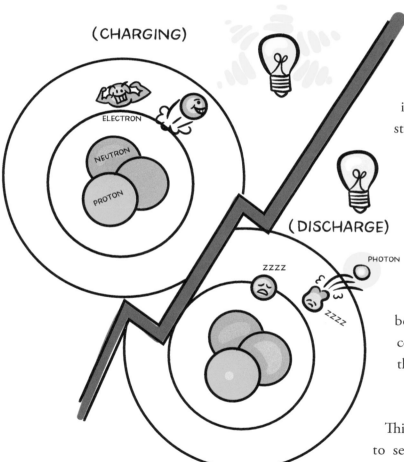

Remember standing on a teetering stack of books precariously positioned in the middle of your slumping mattress so that you could stick cheap, plastic, glow-in-the-dark stars to your popcorn ceiling?

Admit it, you burgeoning interior designer: ya did it. We all did.

The thing about this particular question is that it begs different responses because there are actually a couple of different mechanisms that yield glow. So! Let's do it.

GLOW VIA FLUORESCENCE: This is the paint you're likely used to seeing on psychedelic posters behind the counter of smoke shops. The paint absorbs light from the ultraviolet (or UV, as you've heard it called) end of the electromagnetic spectrum. Indoors, black lights are a fantastic artificial source of UV. As the paint pigment soaks in the energy, it charges up the electrons within. These hyperactive electrons then subsequently burp out that excess energy in the form of visible light. The glow is short-lived and can usually only be seen under UV sources.

GLOW VIA PHOSPHORESCENCE: This is the traditional glow-in-the-dark paint you probably recalled to mind when you read the subject of this section. This mechanism of glow is similar to fluorescence, but the paint contains compounds called "phosphors." These compounds when light-charged still release energy in the form of light, but it manifests as a long-sustained, pale-green glow. Unlike fluorescence, this type of glow will continue long after the light source has been removed.

This page has now provided you with the scientific rationale to describe most electronic music festivals.

HOW DOES RUBBING ALCOHOL KILL GERMS?

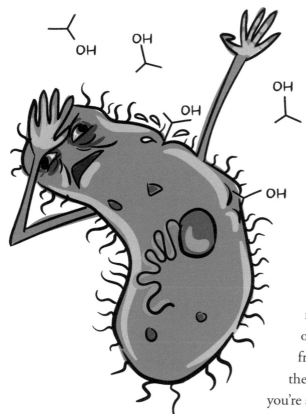

During the first year of the SARS-CoV-2 pandemic, it was damn near impossible to keep sanitizing products on the shelves of stores. In the United States alone, hand sanitizer sales soared to a hefty $1.5 billion. However, despite sanitizer's now routine utilization, most people are unaware of how alcohol-based products work to keep us from spreading disease. Which is a shame because the mechanism is delightfully interesting . . . unless you're a microbe, of course.

The little bottle of sanitizing gel you were given at your last corporate event, conference, or wedding serves to kill bacteria and viruses by essentially melting them from the inside.

There are typically two different alcohol molecules leveraged in sanitizing products: ethyl alcohol or isopropyl alcohol. They differ slightly in structure, but they work fundamentally the same way. Both alcohol molecules are wildly proficient at the art of inserting themselves into the membranes of many kinds of microbes. Once inserted, they act like the molecular version of a splitting maul, disrupting the tight packing of lipid molecules that form the protective coating of the microbe. This compromise of membrane structural integrity allows for the passage of additional alcohol molecules into the microbe. The alcohol then serves to interrupt the internal, functional proteins of the microbe, changing their structure and rendering them completely useless. This warping and melting of proteins is called "denaturing," and it ultimately becomes the death knell for the rapidly disintegrating bacterium or viral particle.

Some of you may be wondering about the hand rubbing associated with alcohol-based sanitizing products—is it even necessary? Funny enough, it provides a *massive* boost to the efficacy of the process. The vigorous rubbing generates mechanical forces that can help the alcohol molecules to pop microbes open a bit more effectively. So make sure you do the whole "rub your hands together for the length of time it takes to sing 'Happy Birthday' twice" thing. Also, make sure you sing it out loud—this has nothing to do with the clinical efficacy itself, but it's highly effective in confusing anyone who may be within earshot.

WHAT IS FIRE MADE OUT OF?

Throughout human history, our species has always maintained a deep obsession with fire. From grunting at each other around Pleistocene-era campfires to being mesmerized by Kurt Russell's performance in *Backdraft*, fire has occupied a central role in our communities and our imaginations.

But *fire* is a broad term that describes a combustive fuel source, a chemical reaction, and the resultant products they create together. I don't think you meant to ask about that, though—I think you meant to ask about the flame itself. Which is way cooler to talk about. With that: the flame is not really a substance. The flame is an ongoing chemical conversion of the fuel source by oxidative gases, yielding thermal energy and light.

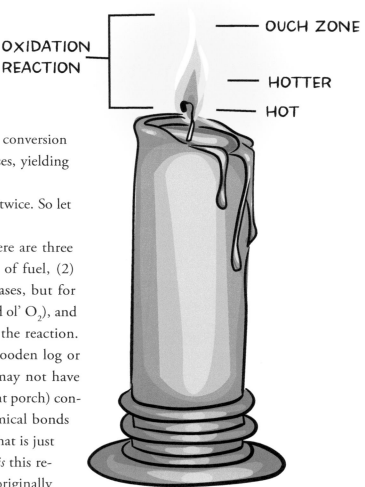

You probably had to read that twice. So let me explain.

For a flame to be created, there are three basic requirements: (1) a source of fuel, (2) oxygen gas (or other oxidative gases, but for our purposes, we'll stick with good ol' O_2), and (3) a source of heat to kick-start the reaction. The fuel source (whether it's a wooden log or a bag of dog poop you may or may not have placed on your neighbor Ted's front porch) contains chemical bonds. These chemical bonds hold a certain amount of energy that is just itching to be released. The flame *is* this resultant release of energy—stored originally within the chemical bonds between atoms in a campfire log and released as the distinctive heat and light of flames.

So the flame is energy; it is the chemical reaction that takes place between oxygen gas and the molecular bonds within the fuel. While it's a bit too detailed to cover here (and would likely put you to sleep), the reaction involves some bonds breaking, other bonds forming, and electrons being shuttled around.

An interesting thing about flame is that it has discrete layers in which different phases of the chemical reaction take place. If you've ever stared into the flame of a candle, you've actually seen these layers. Closest to the fuel source is the innermost layer of the flame—it is the coolest part and contains a mixture of hot vapor from the fuel source and a little oxygen gas. The middle layer of the flame is where the chemical reaction begins to gain its momentum—it is much brighter and hotter than the inner layer. Finally, the outer layer is where the chemical reaction races at full speed—it is the hottest and thinnest layer of the flame.

So the next time you roast marshmallows, you can thank hot, ongoing electron exchange, courtesy of the highly oxidative gases in our atmosphere that you also breathe.

WHAT IS ACID RAIN?

Acid rain is exactly what it sounds like. There's really no other way to describe the phenomenon. It's like a typical rain shower . . . if you replaced the fresh droplets of water with caustic acids. Unfortunately, it's also predominantly a product of human industry.

It all starts with a sulfur dioxide and/or nitrogen oxide compound. In this case, they're both gases and they're both toxic. These gases are commonly released via the burning of fossil fuels (or by the occasional volcanic eruption) and, with respect to human infrastructure, as a by-product of running large electrical power plants. As the gases are released into the atmosphere, they begin to chemically react with water vapor and an array of oxygen species, eventually creating sulfuric acid and nitric acid droplets. If these airborne bits of acid are blown into a weather system, they will mix with the water precipitates (which can be rain, snow, or even fog) and fall to the ground.

While being exposed to acid rain won't result in holes being seared through your flesh (making you look like a sentient slice of swiss cheese), it is still quite harmful. The exact pH of acid rain varies, but it usually exhibits a similar acidity to that of red wine. While that may not raise any alarms in your mind, this pH level is approximately twenty-five to thirty times more acidic than normal rain. As such, acid rain can ravage the environment, burning through the vegetation of forests and destroying valuable minerals within the soil. Additionally, this acid can be washed into adjacent bodies of water, where it can also have devastating consequences on fragile aquatic ecosystems.

The only real way to address the problem of acid rain is through prevention. This means, for us as a species, moving away from our fossil fuel dependence and utilizing renewable energy sources. Or, ya know, we can also just not be proactive at all and see where that takes us . . .

Spoiler alert: acid rain will be the least of our worries.

WHY DOES ADDING SALT TO A PLUMBING SYSTEM "SOFTEN" THE WATER?

The first time I lived in an area with hard water, I was already an adult. I strolled into a hardware store only to find a monolithic pallet warped under the heft of stacked yellow salt bags. In my ignorance, I just assumed a large proportion of households in the area must have saltwater swimming pools. In hindsight, this made little sense considering the small town was predominantly populated by broke college students. But I digress. I'm also getting ahead of myself.

Hard water is a term used to describe a water supply rich in minerals—specifically the minerals magnesium and calcium. This type of water is usually found in communities that pull from groundwater reserves, wherein the water supply filters through natural limestone or gypsum rock beds, scooping up dissolved minerals as it flows.

While hard water is not necessarily bad for your health, the buildup of these minerals within pipes can contribute to a more frequent need for plumbing repairs, the staining of sinks and tubs, accelerated fading and breakdown of fabrics during wash cycles, and an increase of in-home water utilization. Fortunately, because hard water is a problem of calcium and magnesium ions, we can utilize clever chemistry to help mitigate some of these issues.

Back to those big yellow salt bags I mentioned

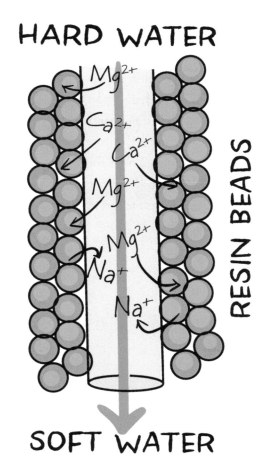

earlier. Salt is an essential ingredient for a chemical process called "ion exchange," which is basically the switcheroo of one charged ion for another similarly charged ion. In the world of correcting hard water, the idea is to capture the positively charged magnesium (Mg^{2+}) and calcium ions (Ca^{2+}) from the hard water and replace them with positively charged sodium ions (Na^+) from the salt. Why? Well, sodium ions in a water supply are *far* less problematic than the alternative.

This process of removing the minerals from hard water is called "softening" the water. Inside a water-softening system is a network of pipes that shuttles the hard water into a tank, where it interfaces with hundreds of thousands of small resin beads. These beads carry a negative electric charge. When the resin is fresh, it is coated in positively charged sodium ions from the salt we talked about. As the hard water is pumped through the beads, the positive charges on the magnesium and calcium ions *also* become attracted to the negatively charged beads. This causes the hard minerals to get pulled from the water as they stick to the resin, which kicks the sodium ions off the beads and into the water (this is the ion exchange part), and the freshly softened water is pumped into your home for usage.

The salt itself is necessary to "recharge" the system. It doesn't take long for the hard water minerals to coat the surface of the resin beads. With no other spots for these hard water ions to bind, the minerals would pass right through the beads and the water would remain hard. So salt water is flushed though the system in order to, in a sense, chemically power wash the hard minerals from the beads, which allows for more ion exchange and for the water-softening process to continue.

WHY ARE SOME ELEMENTS RADIOACTIVE AND OTHERS ARE NOT?

The forces that dictate the dynamics of our universe probabilistically drive the world around us toward a state of energetic stability. This is outlined in the laws of thermodynamics, which collectively describe the natural tendency of any system (whether that be a ball suspended by a spring, a rocket fired into space, or a cup of hot coffee) to eventually reach energetic equilibrium. Like the physical and chemical version of reciting so many "om" incantations, events that transpire within our universe, if unperturbed by extraneous forces, will eventually find balance.

So what the hell does this mean?

IN SHORT FORM: Radioactive elements shed a part of themselves to release excess energy in an effort to reach energetic stability. Nonradioactive elements already exist in an acceptable state of energetic stability.

IN LONG FORM: Matter, as we know it, is made up of atoms. Each atom contains a central nucleus, which is composed of subatomic particles called "protons" and "neutrons." The nucleus is also surrounded by zippy, elementary particles called "electrons." Electrons are negatively charged, protons are positively charged, and neutrons have no charge. To keep this cast of subatomic characters together, a couple of fundamental forces are at work: the strong nuclear force (which glues the protons and neutrons of the nucleus together) and the electrostatic force (which keeps the electrons tethered to their respective nuclei).

Some atoms naturally exist with imbalanced ratios of subatomic particles—too many or too few protons, for instance. This imbalance leads to an asymmetry of atomic forces, and as a result, the atom exists in an energetically excited state. But atoms don't like this—they are happiest if they can reach their zen, their Swiss neutrality, their thirty-minutes-after-taking-nighttime-cough-medicine levels of energy. Radioactive decay is the atomic attempt at correcting these imbalances and achieving the atom's chillest vibes.

Several types of radioactive decay can be employed by atoms. Each of these decay types may release different particles or various flavors of ionizing radiation, dependent upon what the atom needs to adjust to correct its imbalances. Therefore, what makes an element radioactive, and how quickly it decays, is based upon the energetic damage control that needs to take place. By contrast, nonradioactive elements are the Jeff Lebowskis of the periodic table: they don't need to undergo decay—they've already achieved full chill.

You didn't anticipate a simple question about radioactivity evoking deep ponderance of the zen of the universe, did you? Welcome to Science.

WHAT MAKES ICE SLIPPERY?

Water is both omnipresent and irrefutably underappreciated. This little molecule has been integral to both the establishment *and* continued fostering of life on our metallic space rock. Its unique chemical properties make it a formidable solvent and a fantastic medium for thermal regulation. These properties also confer fascinating bonding characteristics to water in its solid state. In fact, when water freezes into ice, it counterintuitively becomes *less* dense than its liquid state. This is precisely why the ice cubes in your iced tea (be it your grandmother's recipe or the Saturday-night Long Island version) float.

What makes ice slippery actually has nothing to do with the solid ice itself. Ice is slippery because of an imperceptibly thin layer of liquid water that coats the surface of the ice. Please note that the usage of the word *imperceptibly* is not an overstatement—the liquid water layer is only nanometers thick, and certainly something you wouldn't be able to distinguish from the ice itself simply by looking at it. Ultimately, this sneaky surface puddle significantly decreases properties of friction on the surface of frozen water. This allows ice skaters to glide effortlessly atop their foot-blades, and it also allows the video you took of your neighbor, tumbling while shoveling his driveway, to go viral on social media.

It took the scientific community quite a while to understand the rationale for the origin of this slick surface liquid. Scientists once thought that things like friction or pressure were responsible for melting the outer layers. While these factors can and do play a role, the liquid remains even when you remove some of these variables—curious findings.

The explanation for this phenomenon is actually associated with the way water molecules bind to one another. When in a solid state, water molecules stick together via hefty chemical bonds called "hydrogen bonds." They link up in coordinated alignment, tightly packing into a tidy lattice. However, the water molecules on the surface of ice aren't packed in by other water molecules above them—they're exposed. This allows for more physical wiggle within the outer row of molecular water, and oftentimes this movement is enough to change the state of water from a sturdy solid into a moving fluid.

The nuances that determine states of matter can be subtle, but it goes something like this: molecules in solid matter don't move much, molecules in liquid matter move a bit more, and molecules in gaseous matter move a lot more. That's it. Nothing really changes with respect to the molecule, apart from how much it moves. Which means, in the case of myself, having not moved much on this particularly lazy Saturday, I would be classified as a very immovable crystalline solid—low energy and limited motion.

COULD YOU ACTUALLY MAKE A DIAMOND OUT OF COAL?

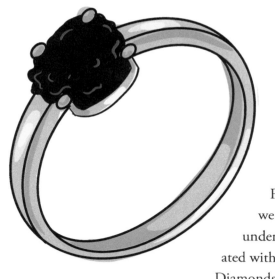

For *any* hypothetical question, it's super important that we understand the foundational concepts of the system under consideration. So let's talk about the processes associated with natural diamond formation on Earth.

Diamonds are made up of densely packed, tightly arranged, covalently bound carbon atoms. To form the immensely strong internal structure of a diamond, each carbon atom is tethered directly to four other carbon atoms in a basic unit that chemists call a "tetrahedral" configuration (the shape of a tiny four-faced pyramid). Tetrahedral units of carbon bind to other tetrahedral units of carbon in one enormous repeating, three-dimensional pattern.

Why do I describe this to you? Only to tell you that a diamond (and other crystalline solids) is super-*duper* strong. Which leads nicely into the discussion of how all these carbon atoms become so closely and tightly mashed together.

In the world of chemistry, atoms prefer to have personal space, for a few reasons. They don't really want to be abutted next to other atoms unless they have a lot of external encouragement. So to squish 10,000,000,000,000,000,000,000 carbon atoms into a single carat of diamond (and, yes, that ridiculous-looking number is an accurate estimate of carbon atoms for that particular weight of diamond—the number is ten sextillion), you need an immense amount of pressure and heat. However, this kind of naturally occurring, pressure-cooker environment does not exist on the surface of our planet, apart from perhaps a meteor impact site. Instead, natural diamonds are manufactured *inside* of the earth, at the level of our planet's mantle, where temperatures can exceed 6,500°F (3,600°C) with pulverizing pressures as high as 200,000 psi.

The next thing to understand is that the materials within the mantle are billions of years old.

This means the stones in your sparkling jewelry are also billions of years old. While diamonds may be found in layers of rock dated to be much younger, it's because they've been transplanted from the mantle, guided by the upward force of deep volcanic activity. This is where we look for diamonds today—in areas that exhibit the remnant pipes of ancient eruptions.

Now, as to the somewhat common assumption that coal can be crushed into a diamond . . . as far as we know, this is actually *false*. Here are a handful of critical reasons why.

MATERIALS: Diamonds are made of a latticework of *pure* carbon extracted from rock sources deep within the earth. Coal, on the other hand, is made of hydrocarbon molecules from ancient plant life that died and became submerged under primordial swamps—superficial Earth-surface stuff and quite an *impure* source of carbon.

MECHANISM: The crystalline structure of diamonds suggests that carbon isn't simply crushed down to yield a diamond. Instead, diamonds are *grown* very slowly in a process that may take billions of years.

TIMING: There is no substantial evidence to suggest that coal has *ever* yielded diamonds. In fact, most of the naturally occurring diamonds on this planet began their formation billions of years in the past. This timeline puts diamond creation well before plant life ever existed, making coal an implausible source of carbon.

MODERN TECHNIQUES: Scientists can now replicate the temperature and pressure necessary to make diamond material. As such, the burgeoning industry of lab-created diamonds has seen tremendous advances. However, lab protocols do not include crushing rocks into diamonds, but rather they follow a process of growing diamonds, not unlike natural mechanisms.

So, when taken together, between the impurities in coal, the lack of evidence to suggest that coal has ever been utilized for natural diamond creation, and the processes of diamond creation being closer to growing as opposed to crushing a rock until it becomes a slightly different rock . . . I think we can safely close this chapter. Make sure you file this away under "Highly Scientifically Unlikely, Yet Still Somehow Widespread Information."

WHY IS CARBON MONOXIDE SO DANGEROUS?

Carbon monoxide is a silent assailant. This gas is colorless, odorless, and reportedly tasteless (although, I'm not sure who is responsible for tasting gases—Senior Gas Taster seems like a tough job for human resources to fill). The difficulty in its detection is made worse by the fact that prolonged inhalation of this molecule can be lethal to humans and animals. In the United States alone, approximately 50,000 people receive emergency treatment for carbon monoxide–related complications annually.

When a complete combustion reaction is generated from carbon fuel sources, one of the by-products is carbon dioxide gas. By contrast, when an *incomplete* combustion reaction occurs, carbon *monoxide*

is evolved. Incomplete combustion typically occurs when there is a lack of oxygen in the combustion environment. We can see the difference between these gases with a quick examination of their respective molecular formulas: complete combustion yields carbon dioxide (CO_2, which is one carbon atom linked to two oxygen atoms); incomplete combustion yields carbon monoxide (CO, which is one carbon atom linked to one oxygen atom). The carbon monoxide gas is missing an oxygen atom compared to its counterpart. Which should make sense because of the whole lack-of-oxygen-during-its-creation thing.

There is a multitude of man-made sources that can release carbon monoxide, from automobile engines and portable generators to faulty heaters and boilers. So exposure to this volatile gas is commonly reported from inside the home.

Carbon monoxide poisoning is associated with several physiologic disruptions within the human body. One of the primary dangers is related to carbon monoxide's strong chemical attraction to a protein called "hemoglobin." Hemoglobin is located inside red blood cells and is responsible for facilitating the transport of oxygen throughout the human body. When carbon monoxide is inhaled, it enters the bloodstream, tightly locks onto these hemoglobin molecules, and disrupts their ability to bind with oxygen. This molecular displacement decreases the amount of oxygen that can be delivered to important tissues. As the levels of blood-oxygen saturation dwindle, critical areas like the brain become hypoxic and can suffer devastating tissue injury. For this reason, victims that survive significant carbon monoxide poisoning may experience neurological impairment.

To guard yourself and your loved ones from carbon monoxide poisoning, it's important to know the symptoms. These include headache, fatigue, dizziness, altered mental state, nausea, and shortness of breath. It's also important not to keep forgetting to change the batteries in your carbon monoxide alarms. Just like you continue to neglect the batteries in your smoke alarms . . . I can hear their shrill chirps from here, you animals.

HOW DOES FLUORIDE IN TOOTHPASTE HELP PREVENT CAVITIES?

Fluoride has been heralded as the patron saint of dental health since the early twentieth century. However, somewhat recently, the clinical efficacy of fortifying food and water supplies with fluoride has become hotly contested. Health complications associated with excess fluoride ingestion are the reason for the aggressive warnings (typically in a bold typeface) on toothpaste tubes that read Do Not Swallow. As concerns about fluoride toxicity mount, several countries have stopped fluoridation of public water supplies altogether.

But we can let the World Health Organization duke it out with scientists from the international community about systemic fluoride. For the purposes of this entry, *topical* applications of fluoride to teeth are still deemed safe and are not currently in the crosshairs of controversy. It turns out, fluoride packs a pretty big chemical punch when it comes to cavity prevention.

Fluoride can prevent cavities via a couple of interesting mechanisms. But to understand these, you need to know a little bit about the formation of cavities themselves. You may have grown up with parents who advised you not to eat too much candy, or you'd get cavities. While they weren't wrong, the sugar itself is not what causes the cavity—sugar actually feeds the organisms that cause the cavity.

Dental cavities begin as a localized weakening of the super hard, protective enamel that coats the tooth. If unchecked, the breakdown of hard tissues may eventually progress to tooth decay. This damage is caused by acid produced by resident bacteria inside of your mouth. Bacteria like *Streptococcus mutans* and certain *Lactobacillus* strains will gorge themselves on sugars that remain in your mouth, post–candy bar binge. They use the sugar molecules to do two things: (1) encapsulate themselves in a sticky layer that helps them adhere to the surface of the tooth (known as dental plaque) and (2) produce lactic acid. Over time, the acids from the bacteria will demineralize and destroy the hard layers of the tooth, leading to sensitivity, pain, and permanent dental damage.

The application of fluoride to the teeth and gums can slow the bacterial activity that leads to cavity formation. For instance, fluoride adversely affects bacteria by slipping inside the cell and disabling enzymes that are critical to their metabolism and survival. This decreases the number of viable bacteria inside of your mouth and, by extension, the amount of acid production on the surface of the teeth. Fluoride can also help rectify existing damage by bolstering the structural support of the tooth itself. When fluoride enters the hard tissues, it drags calcium and phosphate ions in with it. This results in tooth remineralization and surface rebuilding in areas that may have sustained previous damage. Repair processes, which involve the restoration of these critical minerals, also serve to make teeth more resilient to future breakdown.

For my part, I am a chronic overbrusher. This means that I—performing full Lady Macbeth scrub sessions in my mouth—have gleaming white teeth . . . that also happen to exhibit a significant amount of enamel abrasion and associated temperature sensitivity.

Ah, the price we pay to have teeth that look like peppermint Chiclets.

HOW DO BATTERIES STORE ELECTRICITY?

To understand how batteries function, you need to know a little bit about how electricity itself works.

Whether it's coursing through the filament of an illuminated light bulb or moving the mechanical arms of a chintzy, overpriced robot you purchased for your niece's birthday, electrical current is simply the flow of electrons. That's it.

The wiring in your home acts as an electron transport conduit from large city generators. The wires themselves flow with a veritable river of electrons, powering all the gadgets and doodads you tap into it. But batteries are small, compact, and certainly not docked into any kind of large city grid. So how are they able to provide electrical power? Well, this all comes down to the conversion of chemical energy into electrical energy.

While batteries come in a wide variety of shapes and sizes, most batteries house two essential components for a successful round of electron harvesting: a couple of different metals and a soup of electrolytes. Different metals make up the cathode (connected to the positive terminal of the battery) and anode (connected to the negative terminal of the battery). The anode metal will release electrons; the cathode metal will accept those electrons. This one-way exchange of electrons between the two metals is the mechanism of electron flow, or electrical current. However, the cathode and anode are separated by a barrier within the battery. It isn't until you insert batteries into a device, power it on, and complete the circuit through the device's wiring that the anode and cathode have a clear road between them to undergo electron exchange.

So what is the electrolyte soup that we speak of? Well, this liquid, paste, or semisolid is what chemically balances the charges at the terminals and what keeps the electrons flowing. It goes like this: The anode and cathode are both partially submerged in the electrolytic solution. As they undergo their exchange of negatively charged electrons, the electrolyte provides a medium for *positively* charged ions to also move. As the anode releases electrons through the circuit wire, it casts off a proportional number of positive ions into the electrolyte. As the cathode receives electrons through the circuit wire, it picks up a proportional number of positive ions from the electrolyte. If each terminal maintains this balance, the electrochemical potential is maintained, and the escaping electrons can continue to flow through the wire and provide you with electrical current—*boom!* You're in business!

To give you a summary (TL;DR, as they say): Electrons contained in the anode metal are itching to leave. The cathode, on the other hand, is anxiously ready to accept electrons. When a wire connects the anode to the cathode, electrons will flow between them and create an electrical current. However, without the electrolyte solution, the current flow would be short-lived. The electrolyte solution provides a repository of positive ions to maintain the balance of charge at both terminals, which prolongs the release of electron flow and ensures that your flashlight won't immediately run out of power (which is especially helpful if you have to go down into your creepy basement).

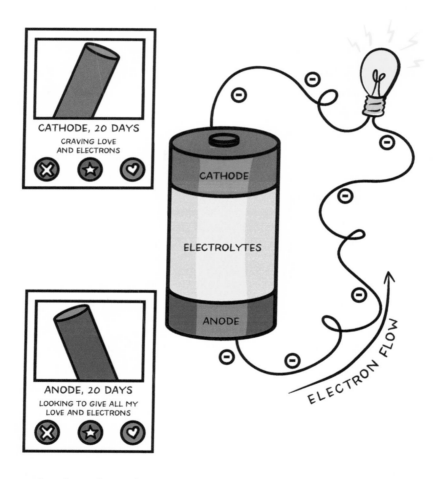

As you can see, the electrolyte solution is essential to a battery's performance. Kind of like the utility of Gatorade to an athlete. But I strongly advise against drinking battery fluid in lieu of a lemon-lime sports drink.

WHAT IS DRY ICE?

It may bake your noodle a bit when I tell you that dry ice is actually frozen carbon dioxide. When you think of carbon dioxide, you likely think of it in its gaseous form. Which is completely understandable because the temperature and pressures of our surface-level, earthling environment lend to carbon dioxide being in its most ethereal state of existence.

Carbon dioxide is an interesting molecule. At regular atmospheric pressure and plain ol' room temperature, it undergoes a transition called "sublimation." This is a process whereby solid matter will convert directly into a gas, completely skipping the intermediate liquid phase. When you see dry ice giving off those thick plumes of vapor, it's because the solid ice is absorbing enough energy from the heat in the room for its molecules to get super excited, break free from the solid block, and shoot off into the atmosphere.

Due to these unique properties, in order for carbon dioxide to maintain a solid state, it needs to be kept at a chilly –109°F (–78.3°C). This is why dry ice comes in handy for keeping food (or even human tissues) colder for much longer than water ice. Its easy sublimation is also super handy for fog machine fuel, helping DJs to bring a little extra magic to the dance floor since its discovery in 1835.

WHY DOES IRON RUST, BUT GOLD DOES NOT?

Nonferrous (or noniron) metals, like gold or silver or platinum, will not rust. If you find that they do, then you should consider having a very stern discussion with your jeweler.

The reason iron rusts is due to simple chemistry. As with any other chemical reaction, the rusting mechanism will not happen without the appropriate ingredients. Nonferrous metals are *not* the appropriate ingredients.

Rusting is an oxidation reaction that occurs between the oxygen in our atmosphere and ferrous metals. During this reaction, atmospheric oxygen—catalyzed by moisture in the air—attempts to steal electrons from the candidate metal material. In doing so, oxygen atoms become bound to iron atoms, forming compounds called "iron oxides." Over time, these iron oxides accumulate as a scaly rust layer on the surface of the metal, weakening the existing iron-to-iron bonds. With enough oxidation and enough rusting, the metal will become brittle and turn into an orange pile of flaky powder.

Other metals, like gold, do not rust or oxidize because they do not have available electrons that are easy for oxygen to lock onto. All chemical bonding is a product of electron exchange and sharing—this is the basis for reactivity. The way an atom is configured and where its electrons exist in space determine how reactive it will be. In the case of metals, some have a reactive configuration with electrons that can be poached (iron); others keep their electrons locked in and inaccessible to oxidants (nonferrous precious metals).

Fun historical fact for you: the Statue of Liberty is copper. When it was originally gifted to the United States, it looked like a freshly pressed penny—she was a rich hue of gleaming orange. Over time, the salty sea air oxidized Lady Liberty, coating her in a layer of copper oxides. And thus she stands today: a green monolithic woman—our nation's She-Hulk.

WHY DOES HOT, BOILING WATER SHATTER A COLD DRINKING GLASS?

Thermal shock (surprisingly not the name of a 1990s-era thrash band) is a transient, deforming force that is applied to materials via drastic temperature changes. It is the reason that dropping ice cubes into a glass of room-temperature water makes the cubes crack. It is also the reason that—as you may have learned the hard, messy way—pouring boiling water into a glass can result in a small eruption of glass shards.

To understand this, we need to take a closer look at what is occurring on the atomic level. Matter is *moving*—gases, liquids, and even solids exhibit motion on the smallest of scales. For instance, if you were to examine the water molecules inside of solid ice, you would see them vibrating. Water molecules in liquid water are meandering past one another, and water molecules in water vapor are zipping around through the atmosphere. As such, the primary determining factor of a state of matter is the kinetic energy associated with its molecules. For the most part, lots of energy = fluids (gases/liquids), and low energy = solids.

When heat is rapidly applied to a material like glass, there is a massive pumping of energy into its molecules. This energy burst causes the molecules to exhibit more motion and occupy more space, which results in the physical expansion of the glass itself. If this deformation occurs rapidly, stress is put on the intermolecular structure of the glass, which can exceed the limits of tolerable strain. Ultimately, boiling water may add enough energy and provide enough strain to cause total structural failure and a sharp shattering effect.

CHEMISTRY

When I studied at Harvard, I was a transplanted native from Southern California, completely naive to *real* cold. I vividly remember a particularly frigid winter morning when I could just about see my breath inside my drafty apartment. I had boiled water to steep cheap green-tea sachets inside a thick mason jar. Craning over the stove, I sleepily tipped my kettle into the glass and was met with a dense *POP!*, followed by a deluge of steaming water that poured over the face of my oven and onto the floor. Stunned, I stood blinking at the mess. But I'm not ashamed to admit that I let my frozen feet linger in the warm puddle for several minutes before wiping it up.

WHY DOES COOKING AN EGG YOLK MAKE IT SOLID?

In nature, particularly in a biological context, the shape of macromolecules (large polymers made from building blocks like carbohydrates and proteins) is intrinsically linked to their ability to correctly function. These shapes—whether curled, swirled, kinked, pleated, folded, or knotty—are all determined by chemical bonding. Each tertiary or quaternary chemical bond acts like a molecular safety pin, keeping specific sections of a molecule folded in place or adhering one molecule to another.

Proteins are incredibly reliant on structure to determine what they do, what they interact with, and what they look like. The inside of an egg harbors a large proportion of proteins in both the egg

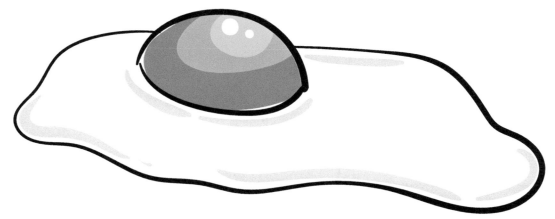

white and the egg yolk (or "egg yellow," as I call it). The structure of the proteins inside the egg white is fundamentally different from the protein structure inside the egg yolk. However, the commonality they share is that both types can be deformed by the application of heat.

The deformation of protein shape and structure is formally called "denaturing," and it is a product of the breakdown of those safety-pin chemical bonds that keep everything neatly in place. Once those bonds are lost, the intended shape of the protein begins to fall apart, possibly binding with other unintended molecules and losing its original properties. So here's how heat changes egg proteins:

EGG WHITE: The proteins of an egg white normally exist in a globular configuration. This is a shape wherein the protein strand folds back and links into itself so many times that it ends up looking more like a knotty sphere (or headphone wires that got tangled, even though you only had them in your pocket for, like, ten minutes). When heat is applied to these proteins, the links that bind the clumpy proteins together begin to break, and the structure loosens up. The continued application of heat provides more energy to the molecules, and they eventually begin to form strong bonds with neighboring proteins. This squeezes water molecules out of the egg white, and the resultant, tightly adhered protein matrix begins to look rubbery and white.

EGG YOLK: The yolk of the egg is meant to provide nutrition to the growing embryo and thus contains an emulsion of both fats and proteins. Similar to the chemical deforming of egg white proteins, hot yolk proteins will also lose their intramolecular bonding, warp out of shape, and aggregate together. The hotter the applied temperature, the tighter the proteins lock together, the more water is extruded, and the grainier and drier the yolk becomes.

There's definitely a corny, inspirational meme in here somewhere. Something to the effect of, "I'm like an egg, bro. Apply a little heat to my life, and I come out harder than when I started." Yep. That's it—that's the new Live, Laugh, Love wall mantra.

I'm actually chuckling right now.

WHAT IS BPA IN PLASTIC, AND WHY IS IT HARMFUL?

You've probably seen the stickers slapped on the sides of plastic food or beverage containers that read, "BPA-free." You may have also read those stickers and thought, "Oh, cool . . . I guess?" So, let's bring some context into these reassuring marketing messages.

The initialism BPA stands for a chemical called "bisphenol A." This compound acts as a structural reinforcing agent for hard plastics and epoxy resins, and its utilization is widespread. Currently, you can find BPA incorporated into the plastics of reusable food containers, water bottles, baby bottles, and the resin lining of aluminum cans.

The likely mode of direct human contact with BPA is through ingestion. It has been found that this compound can actually integrate into the consumable products contained within these BPA-plastic containers. The degree to which BPA leaches into foods and beverages can also increase as temperatures of the container rise, making exposure more likely when these containers get microwaved or sit in hot vehicles.

Between 2003 and 2004, a study was conducted by the Centers for Disease Control and Prevention

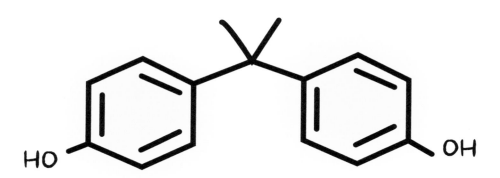

to understand possible BPA exposure rates within the United States. This evaluation included data from more than 2,500 human subjects. The findings of the study indicated that detectable BPA was found in more than 90 percent of urine samples collected. A resounding majority.[8]

Here's the problem with widespread exposure to this little compound: it may be associated with adverse health effects. Reviews of epidemiologic studies indicate that BPA exposure may be associated with a battery of poor health outcomes in infants, children, and adults. For instance, some studies indicate a possible association between BPA and (1) unfavorable fertility metrics in women, (2) decreased sexual performance in men, (3) low birth weight in newborns of exposed parents, (4) increased incidence of poor behavioral effects in children, (5) increased risk of type 2 diabetes in adults, and (6) increased incidence of coronary artery disease in adults.[9]

The safety of BPA is still being explored by intensive clinical studies managed by public health organizations. So, while we don't yet have a final ruling on tolerable levels of BPA or its ultimate safety, the corroboration of poor outcomes in published studies is beginning to mount. Which is why I carry a gallon-sized, reusable, stainless-steel water jug to the gym: it's BPA-free, and the added heft of the steel makes it a decent self-defense weapon.

HOW IS THE PERIODIC TABLE ORGANIZED?

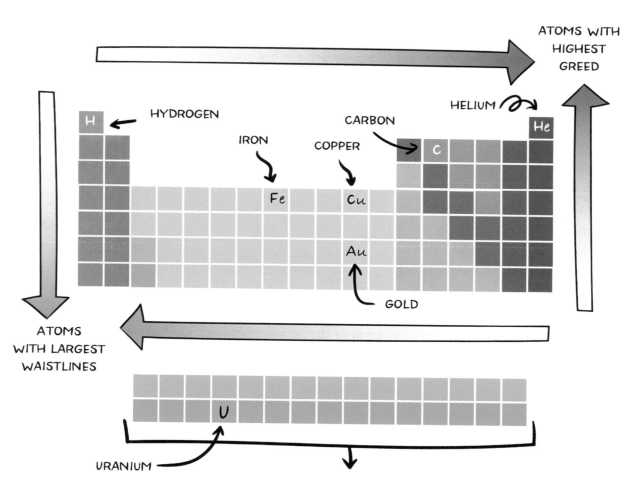

I'll be perfectly honest with you, dear reader, I'm surprised you asked this question. Of all the things—biological, chemical, quantum, or otherwise—that you could have asked, you chose to learn about the rationale for the tabular organization of elements. That is so boring, and I absolutely *love* you for that.

The periodic table is the product of multiple scientists and many, many, *many* years of diligent atomic analysis. In the late 1700s, a chemist named Antoine Lavoisier began the quest for organization by lumping elements into categories described as simply "metals" and "nonmetals"—basic groundwork owing to our lack of knowledge at the time. About one hundred years later, another chemist, Dmitri Mendeleev, took a more granular approach and began to organize the elements according to their atomic weight. However, his stratum of elements left unexplained gaps in the table. Finally, in 1913, with a better understanding of subatomic particle theory, physicist Henry Moseley arranged the elements by the number of protons in their respective nuclei, otherwise known as the "atomic number." This is the order in which the elements appear today.

Currently, the periodic table is organized by periods (rows) and groups (columns). Periods contain metals on the left and nonmetals on the right; groups contain elements that are lumped together for being chemically similar (for instance, a group called the "noble gases" includes gases that are all inert and generally nonreactive). As you read the periodic table, the size of atoms for each element increases from right to left and from top to bottom—this results in the largest elements being situated in the lower-left corner, and the smallest elements in the upper-right corner. Additionally, how strongly an element attracts electrons increases as you go from left to right and from bottom to top—this results in the strongest electron attractors being in the upper-right corner.

The wonderful thing about the layout of the periodic table is it can be used to do things like predict reactivity between atoms or determine the products of nuclear fusion and nuclear fission. A not-so-wonderful thing about the layout of the periodic table is that most general chemistry professors will test your memorization of elemental organization. This leads to you and your classmates making an array of ridiculous mnemonics and songs in an effort to regurgitate the order of elements for your midterm exam. But if you were to ask me to sing them now? It would take me several stiff drinks to open up those ancient wounds.

HOW DO THEY MANUFACTURE FIREWORKS TO BE DIFFERENT COLORS?

I don't know which surprises me more: the fact that Americans annually spend approximately $1 billion on Fourth of July fireworks, or the fact that I wasn't surprised when I read that statistic.

Early firework technology was developed sometime between the seventh and tenth centuries under the Tang dynasty in China. It began as an early gunpowder tincture stuffed into bamboo tubes and set ablaze. Since that time, pyrotechnics have advanced considerably to provide us with smokers, sparklers, poppers, and deafeningly loud boom-boom sticks.

The aesthetic appeal of fireworks is achieved through the manipulation of the temperature levels of their combustion and—importantly—through the chemical constituents contained within the firework fuel mixture. During the explosion of the firework, these chemical compounds become superheated. This heat energy excites the electrons of the firework chemicals, which forces the electrons to change position (the more excited they become, the further from their parent nucleus they get). But electrons don't like to stay in this excited position for long, and when they bounce back to their original atomic homestead, they shed their excess energy in the form of light. The wavelengths of this light vary, based on the chemicals they come from, and these different wavelengths correspond to different colors on the visible light spectrum.

While the firework mixture may differ from manufacturer to manufacturer, the elements that produce color are always consistent. Here's a quick breakdown for you, stratified by color:

- BLUE: **COPPER COMPOUNDS**
- YELLOW: **SODIUM COMPOUNDS**
- GREEN: **BARIUM COMPOUNDS**
- RED: **STRONTIUM COMPOUNDS**
- ORANGE: **CALCIUM COMPOUNDS**
- SILVER: **ALUMINUM OR MAGNESIUM COMPOUNDS**

Now that you have a better understanding of the color origins of your favorite Fourth of July explosion display, I think you can use it to your advantage. Make sure, in the most romantic of moments, that you gaze into your partner's eyes and say, "The aerial display of strontium excitation is so beautiful tonight."

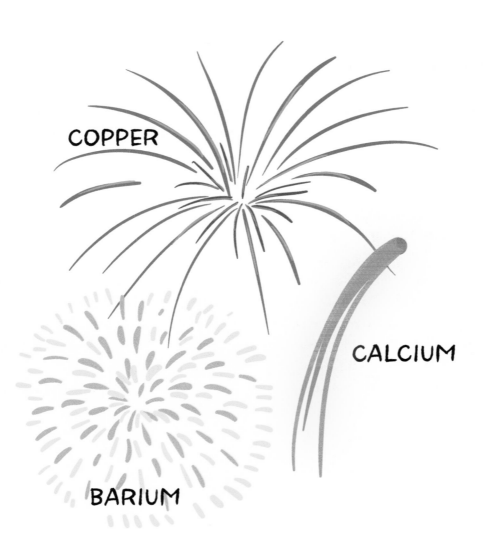

PHYSICS

"IN SCIENCE THERE IS ONLY PHYSICS; ALL THE REST IS STAMP COLLECTING."
—LORD KELVIN, THE PHYSICIST WHO FORMULATED
THE ABSOLUTE TEMPERATURE SCALE

HOW IS A 735,000-POUND AIRPLANE ABLE TO FLY?

My father was a career pilot—both military and commercial—so I've been a passenger on literally *hundreds* of flights. As such, I've never quite understood why some people experience crippling anxiety about flying. I mean, you're bolted to the inside of an aluminum tube carrying 63,000 gallons of highly combustible jet fuel six and a half miles above the ground, while careening through the atmosphere at almost seven hundred miles per hour—what's so scary about that?

(Obvious sarcasm is, hopefully, obvious.)

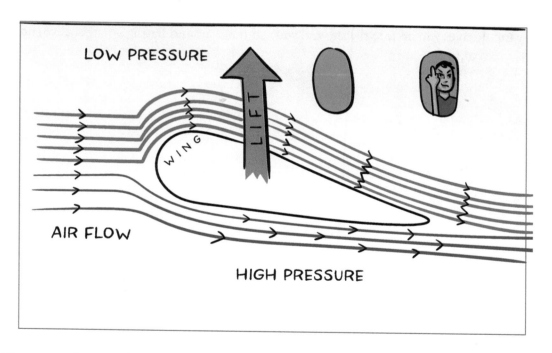

The reason that a girthy Boeing 747 can heave itself into the air is due to the manipulation of air pressure around the wing. You see, flight is made possible by vastly important physical phenomenon described by the field of fluid dynamics. These concepts are riddled with mathematical equations elucidating the relationship between how fast a fluid is moving, the nature of its movement, and the resultant pressure that it would theoretically exert on the area adjacent to it. The air in our atmosphere is classified as a fluid, and in a nutshell, the math shows that the faster a fluid moves, the less pressure it exerts (for those curious, this is part of what we call "Bernoulli's principle").

The shape of an airplane wing is engineered to manipulate the speed and trajectory of air rushing around it. To provide lift, the wing must create an area of pressure imbalance: more pressure underneath the wing, with less pressure above the wing, so that the overall resultant force is upward. Engineers accomplish this feat by crafting wings that direct the air to travel faster above the wing and slower below. When the pilots hit the throttle and accelerate down the runway, they are increasing the disparity in these pressure dynamics. Eventually, the upward force of lift will become stronger than the weight of the plane, and—with the help of various flaps and other clever gizmos—it becomes airborne.

Imagine traveling back to antiquity and trying to explain modern flight to an ancient civilization. "So, there's this big . . . ummm . . . chariot? It carries people, like a chariot, but it's built like a bird. But it's also made out of metal. And it doesn't move by horses—it moves because of . . . well . . . basically fire. It takes you up into the sky and you can travel around Earth. What? . . . No, no. Earth is not flat . . ." Wild.

WHY DOES ICE FLOAT?

I could end this entry right now by telling you that ice floats on liquid water because it's less dense. Done. But, dear reader, I know you probably woke up this morning and thought *extensively* about what intermolecular dynamics look like within different states of water. Therefore, in order to not fundamentally disappoint you, allow me to wax poetic.

If you have already read the entry regarding thermal shock and shattering glass, then you have a decent understanding of the way molecules move within the different states of matter. If you *skipped* that entry and came directly to the physics section, well, you're reckless (I respect that). Here's a brief rehashing of those concepts.

The molecules within matter are in motion. Liquid? Moving. Gas? Moving. Solid?? Oh, yes—still moving. While the constituents of all matter are moving, the relative motion differs per state: gaseous molecules are very zippy, liquid molecules slink and slide past one another, and solid molecules vibrate in place. Now, to the density part.

Water is a unique substance because its solid state is *less* dense than its liquid state. Hence, ice can float. This disparity in densities is attributed to how water molecules bind to one another. When water is solid, it establishes a rigid lattice structure. Based on charges within each molecule and the bonds they form with one another, they need to be aligned in a particular orientation to be static and held in place. However, this orientation has a propensity to leave open space throughout the structure.

On the other hand, liquid water molecules undergo a cycle of continuous bond formation and subsequent bond breakage. This turnover allows for closer interaction between the molecules as they glide past each other and rotate into place to facilitate the transient bonding. This lends to far less average open space in the structure of liquid water when compared to solid water.

So there you have it: the molecules within solid water are not as densely packed as their liquid counterparts. Due to this, the solid state of water gets displaced by liquid water, and ice floats to the top. Which works out pretty well for our planet. Ice is actually a great insulator when it freezes over the top of bodies of water, supporting aquatic life by keeping it cozy and warm in frigid conditions. That is, of course, if you consider 40°F (4°C) to be "cozy and warm."

WHAT IS QUICKSAND?

What is quicksand? According to the magic of cinema, quicksand is a really great way to trap bad guys, should you find yourself being chased through a jungle.

The best way to describe the physics of quicksand is to compare it to a crumbling house of cards. If you've ever found yourself as a young child alone in your room for hours, building teetering structures out of an old Bicycle card deck you found in the garage shortly after your parents' divorce, when you didn't really have friends, because you were the weird nerdy kid who liked dinosaurs *too* much . . .

. . . sorry, I think I just got lost in a personal flashback . . .

Anyway, if you've ever tried to stack cards like that, then you understand, intimately, how the slightest imbalance can bring your tedious work crashing down into a most pathetic pile. Quicksand is like that. But the thin cards are replaced by oddly shaped grains of sand or other mixed substrates.

On a tiny level, these weird shapes rakishly lean against one another, leaving proportionally large gaps of air or water in between. When undisturbed, they can exist like this into perpetuity. But! With some application of force or pressure from the foot of a very unlucky traveler, this sand abruptly caves in, grain by grain. The short-lived soft till collapses inward around the foot (or arm, or head, or whatever—I don't presume to know how you travel), and the sand packs in tight. The more thrashing you do, the more of this earthy card deck you drag down around you, effectively locking you in. To create quicksand, a source of underground water leaches into the uppermost layers of soil where it creates pockets among the grains. So the corny chase scenes in movies are at least somewhat accurate: you are far more likely to find quicksand in a wet jungle as opposed to a desert.

I went ahead and looked up the statistics for you. Surprisingly, quicksand seems to be a highly unlikely harbinger of death. The only real statistic I found was a newspaper clipping from Texas, which stated that one man, out of fifty-two people reported to have been stuck in quicksand, had perished. The article went on to say that the cause of death was likely from exposure to the elements. Which makes me feel like we're not *really* doing our part to thwart pursuant villains.

WHY IS THE SKY BLUE?

Every once in a while, I am asked to speak in front of a middle school or high school class. I get the opportunity to convince the students that science is cool, and they get to avoid classwork for an hour by turning me into a human Google—it's win-win for everyone. The sky question invariably comes up. Every time. It has also been lobbed at me by several adults as well.

The white light that makes its way to us from the sun is not really white at all. In fact, white light is a combination of the colors that we perceive in the visible part of the electromagnetic spectrum. Listed from longest wavelength to shortest, the colors are red, orange, yellow, green, blue, indigo, and violet. This hidden color combination is precisely why raindrops refract sunlight and reveal vibrant rainbows. The light beams bend as they pass through the water droplets, which effectively break apart the primary beam of light into its constituent wavelengths.

However, raindrops aren't the only medium that can influence the behavior of light. In fact, the molecules that make up our atmosphere (including various gases, dusts, and particulate matter) can also change the directionality of light beams. The reason our sky is blue is that the short wavelength associated with the blue part of the visible light spectrum is extremely prone to scattering as it travels through our atmosphere. So as the white light of the sun makes its way to Earth's surface, the blue wavelength gets teased out by the molecules it encounters and subsequently bounces around in our atmosphere. This scattering effect is called "Rayleigh scattering." We don't see the other colors because they stay relatively combined together and are still perceived by our eyes as white.

This principle is well demonstrated by taking a gander at the skies of other planets as well. Because they bear different atmospheric cocktails than ours, the molecules suspended in their air will affect white light differently. For this reason, Mars has a sort of dusty-orange sky as dictated by a heavy serving of iron oxide dust. Additionally, Saturn's southern hemisphere has been reported to exist beneath a sort of yellowish atmosphere due to the presence of ammonia crystals, while its northern hemisphere was observed by the Cassini spacecraft to be bluish.

It is likely for the best that Earth's skies happen to be blue, though. I think Louis Armstrong's "What a Wonderful World" would have sounded far less endearing with the lyrics, "I see skies of ammonia-laden yellow . . ."

HOW DO MAGNETS WORK?

A musical group—with an exceptional amount of cult popularity—called the Insane Clown Posse was abruptly launched into mainstream culture in 2009 with the release of a song called "Miracles," which I would describe as sort of a rap-rock appreciation ballad for natural phenomena. It was parodied and spoofed and chided. Some of the song's more eloquent lyrics included, "F***ing magnets, how do they work?" Silly as the lyrics are, the mechanisms that drive the forces of a magnet are actually quite complex. So, to the Insane Clown Posse lyricists, Violent J and Shaggy 2 Dope, your point is well taken.

Electromagnetism is one of the four fundamental forces that help us describe the physical nuances of the universe (the three others are the strong nuclear force, the weak nuclear force, and—of course—gravity).

There are a few different types of magnets and ways to generate a magnetic field. Although, I'm pretty sure you're asking about a plain old bar magnet. So let's go with that.

Bar magnets have two distinct poles: a north pole and a south pole. The magnetic force (which is basically just an energy field) emanates outward from the north pole and is drawn back into the south pole. In physics we use something called a "vector field," which looks like a map of a bunch of little arrows, to provide a visualization of the direction and strength of forces in a discrete area. Sidenote: I always thought vector fields looked like an image of tiny wind socks that show you how hard and from which direction the wind is blowing. Anyway, if you were to make a vector field for a bar magnet, you would see that the arrows point outward and away

PHYSICS 101

from the north pole, then loop back and reenter the magnet at the south pole. There can be no north pole without a corresponding south pole, and vice versa. Why do magnets have poles that occur only in pairs? We're not absolutely sure. But it's the way our universe is set up.

The force of magnetism is what you feel when you stick a truck-stop keepsake to your refrigerator—it's that sharp, pulling attraction. This attractive force is determined by the state of the electrons inside the magnet. Electrons, being subatomic particles, can be described by their charge, their mass, and how they move. In general, electrons happily exist in pairs. However, they desperately seek their own unique identities. You see, all electrons have the same mass, the same charge, and can exist together in the same location . . . so, how would you tell a pair of them apart? Well, in the world of physics, the individual electrons, in a pair, can be identified by how they move. We call this their respective "spin." If one electron in the pair spins one way, the other electron *must* spin the other way—this is physical law (described by something known as the "Pauli exclusion principle"). Each spin direction exhibits its own little force, but because the electrons are spinning in opposite directions, the pair of opposing forces usually cancel out. This is how it works for most atoms.

. . . except when it comes to ferromagnetic material . . .

In ferromagnetic materials, like those that are inside of a magnet, atoms have some electrons *without* a paired partner. This means they can spin—and subsequently direct their tiny force—in whichever direction they'd like. What ultimately happens is that all of these unpaired electrons, in all the atoms in a chunk of ferromagnetic material, begin to align their forces and spin in the same direction. These trillions of tiny, individual forces then begin to add up, resulting in one large magnetic force with a very specific direction. This is how the north and south poles of a magnet are manifested—through the alignment and subsequent combined magnetic force of trillions and trillions of unpaired electrons.

Of course, the above explanation would have been likely too wordy for rap lyrics. So, I totally understand why the members of the Insane Clown Posse elected to describe magnets as miraculous.

WHY DON'T HUGE SHIPS SINK, EVEN THOUGH THEY'RE MADE OUT OF METAL?

You can toss a 0.088-ounce penny into a fountain, and it will sink directly to the bottom. But if you release a 1,400-foot-long, 220,000-ton cargo ship into a body of water, it floats like a cork. Witchcraft? No. Clever physics? Absolutely.

The floating of metal ships is made possible by the principles of buoyancy. Buoyancy describes the upward force exerted on objects that are floating upon, or submerged within, fluids. This force directly opposes the weight of the object, which pushes down into the fluid itself. So evaluating an object's capacity to float is simple: we determine if the force upward (buoyancy from the fluid) is stronger than the force downward (weight from the object). If the force of buoyancy is stronger, the object will float; if the force from the weight of the object is stronger, the object will sink.

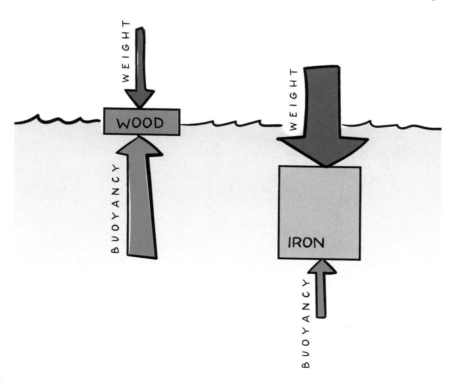

PHYSICS 103

Okay, let's keep unpacking this. We need to understand what determines the strength of each of these forces to know which one wins in a given situation. The force of buoyancy is equal to the weight of total water displaced by the parts of the object that are submerged. Why? Well, this all comes down to the object-fluid surface area interface and the corresponding pressure gradients of the fluid. But when the math shakes out, that upward force is spot-on equal to the weight of the amount of water displaced. This is why ships with more surface area can carry more cargo—the more water they displace, the stronger the upward force becomes.

With respect to the force downward, we have a couple of things to keep in mind. First, the weight of a ship is displaced over a rather large area. Second, the inside of the ship, barring the presence of cargo and personnel, isn't completely solid—it's hollow with a substantial amount of low-density air contained inside. So between the large volume of the ship and the empty space within it, the overall density of the watercraft is quite low compared to the density of the amount of water that would occupy the same amount of space.

When two objects have the same volume, the one with a smaller density will have less mass, and thus its weight will be smaller. Therefore, the relatively low-density ship has a downward force that is *smaller* than the upward force from the relatively high-density water underneath. In this scenario, the ship floats. But if the ship happens to sustain a structural weakening and begins to take on water, its overall density will rapidly increase as it fills. If enough water fills the inside of the ship, its resultant weight will become heavier than the buoyant force keeping it afloat, and the ship will sink.

The old idiom is that "loose lips sink ships," when it's really just a simple imbalance in opposing forces that does the sinking.

WHAT WOULD HAPPEN IF EARTH STOPPED ROTATING?

Rest assured that if Earth were to stop rotating, you would absolutely know it, dear reader. There are a couple of ways that this could go down: a painful, decades-long deceleration ending in a slow stop or a no-holds-barred, instantaneous halt. The latter is a bit more chaotic, so let's talk about that one.

Because Earth's axis of rotation runs through the north and south poles, this makes the equator the longest line of latitude on our planet. Due to its large circumference, it also has the quickest rotational velocity when compared to other latitudes. Currently you can clock Earth's equatorial rotation at approximately 1,000 miles per hour (1,609 kilometers per hour).

At the equator (and elsewhere), if Earth were to suddenly stop all rotation, things that you perceive to be static would be rocketed into violent motion, propelled by the momentum from existing on a previously spinning planet. Cars, trees, people, dogs, skyscrapers, huge rocks, small rocks, park benches, homes, and a whole manner of bric-a-brac would suddenly be jettisoned eastward at 1,000 miles per hour. If you somehow survived careening into, and through, other objects tumbling across the shredded surface of our planet, you'd also need to contend with a host of other terrible circumstances.

The terrestrial part of Earth may no longer be rotating, but that wouldn't halt the atmosphere immediately. So once you skidded to a terribly bumpy stop, you'd be battered by the rush of our atmosphere stripping Earth's surface clean at hundreds of miles per hour. Our oceans, also experiencing the continued momentum from rotation, would slosh onto continents, devastating the landscape and obliterating smaller islands.

Let's say you survived the violent propulsion, then the whipping atmosphere, then the monolithic tsunamis. In the wake of immediate destruction, the Earth you know today would become quite different. The sun would no longer make a daily journey across the sky, which would fundamentally change our climate and our weather patterns. Additionally, the contraction of Earth's bulgy middle—which is due to the centrifugal force from its own rotation—would redistribute large areas of ocean. There's a chance Earth would also lose its protective magnetosphere, leaving our atmosphere vulnerable to being stripped by the solar winds—not unlike the fate we suspect Mars met in the distant past. Oh! There would also likely be a series of cascading geological effects—like massive earthquakes and volcanic eruptions—as the tectonic plates mashed together.

To summarize, if we stopped rotating, there would be a cacophony of crashing and booming, followed by unprecedented winds and flooding, followed then by extreme climate and weather-pattern disruption, followed then by a possible loss of our atmosphere to the ravages of the sun's ionizing radiation.

So, you know, if Earth stopped rotating, it wouldn't necessarily be an ideal situation.

HOW DO I KNOW THAT THE COLOR GREEN THAT I SEE IS THE SAME COLOR GREEN THAT YOU SEE?

Personally, I think this entry evokes one of the most interesting answers in this book. It's one of those things that you'll likely lie in bed in the dark thinking about . . . I do, at least.

Historically, the scientific community had a pervasive notion that, for the most part, we all perceive the same colors. The idea was that our standard set of photoreceptors would all receive colored light in the same way due to the specificity of the wavelengths associated with each color. This would then result in a communication of the same signals to our respective brains. *Poof!* The green you see would look like the green I see because our brains are receiving and processing the colors in a standardized way. Makes biological sense.

However, this idea has recently been called into serious question.

Color vision scientists have not only begun to assert that we *might* see different colors, but that there is a high likelihood that we *probably do*! So while you and I may look at the grass and agree that it's "green," it would only be a common linguistic label. The green grass that I see may appear to be a vibrant sunset orange to you, or when I look at a tomato and say that it's red, you may agree with the name "red," but see my version of a deep blue!

The reason that our perceived colors may be different from one another, even though we have the same sensory hardware, is that the brain seems to have the ability to assign color perceptions on the fly—they're not biologically predetermined. These perceptions may be different for us all, depending on the hue assigned to the visual-sensory stimuli sent to our brains early in development.

In a preclinical study from the University of Washington, a team of scientists performed gene editing on adult male squirrel monkeys to attempt to correct their red-green color-blindness.[10] The findings indicated that twenty weeks after gene editing, red-green color discrimination tests improved significantly in the monkeys evaluated. But the most interesting part is that *all* male squirrel monkeys are born with red-green color-blindness. So, after development, they theoretically have no preexisting neural circuitry to be able to distinguish those colors at all. By default. This means that, during the study, their brains may have utilized *existing* neural connections to assign a color to the new sensory input they were getting. So if they weren't born with the neural hardware to interpret the color red but still passed the tests, what color were they seeing? Evidence like this suggests that the hues we assign to our world may be a beautifully unique, individual experience for us all. While most of us may be able to discriminate between different hues, the actual colors may look quite different from person to person.

As a scientist: this idea is fascinating. As a romantic: this idea is super sad—I want to share a sunset with a loved one and know that we're experiencing the same natural brilliance.

Welcome to my life—moments of blissfully ignorant whimsy tarnished by empirical evidence.

HOW ARE ALL THE COLORS OF THE RAINBOW CONTAINED IN WHITE LIGHT?

I totally understand the undertones of outrage when this question is asked. If you've had a cursory encounter with mixing paint colors, then you know that the combination of even a few of them can result in a drab muddy-brown hue. So it's only natural—based on our firsthand finger-painting adventures—to be confused when we're told that white *light* contains *all* the colors in our perceived visible spectrum. Shouldn't the light, instead, be icky brown?

I find it best to begin with the basics, so let's talk about the nature of white light. The light we perceive is also aptly called "visible light" and is part of the electromagnetic spectrum. It corresponds to wavelengths of radiation that fall in the middle of this spectrum, somewhere between super powerful gamma rays and long lumbering radio waves. Therefore, visible light is simply radiation that we have developed sensory receptors to be able to see. Some animals have gone a step further and can also visualize radiation in the ultraviolet or infrared parts of the electromagnetic spectrum as well.

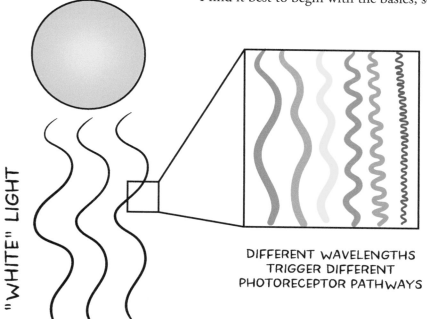

DIFFERENT WAVELENGTHS TRIGGER DIFFERENT PHOTORECEPTOR PATHWAYS

Our range of visible light includes a multitude of frequencies. Human brains have been wired

PHYSICS 109

to interpret each of these frequencies as a distinct hue. So color can be perceived when any of these frequencies enter the eye.

Let's consider a strawberry. The reason we see its bright color is because molecules in the skin of the strawberry have reflected a frequency of visible radiation that corresponds to the color red. The radiation bounces off the strawberry, enters our eye, and triggers a cascade of signaling to the brain, and the brain says, "Yep, that's red." We don't see the strawberry as blue or purple or green because those frequencies were absorbed by the strawberry's skin and, as such, never entered our eye for interpretation by the brain.

White light is what light looks like when none of the frequencies are being reflected or absorbed, and they're all still packaged together. Remember: when photoreceptors of the eye are not being activated, the brain cannot interpret a distinct color. So the light is called "white." But more accurately, it's not really white as much as it is a non-photoreceptor-activating, colorless beam of electromagnetic radiation. If certain wavelengths aren't bouncing off objects, then you won't perceive an associated color.

Now, let's circle back to the frustrations of mixing paints. The reason multiple colors of *paint* will yield icky brown is that multiple combined *pigments* will reflect different frequencies of light into your eyes, simultaneously, for a combined interpretation. And this combined interpretation, with the unchecked paint-mixing bravado of a novice, is usually quite hideous.

ARE THERE MORE ELEMENTS THAT EXIST THAT WE HAVEN'T DISCOVERED?

The current periodic table is an updating, organized repository of the elements we have either discovered naturally or created in a laboratory setting. Not unlike a prized collection of Pokémon cards, we have spent a great deal of time and analysis to organize them in a way that makes the most sense, with special attention paid to how groups of these elements are similarly reactive.

Currently, the elements are organized in ascending order by the number of protons their nuclei contain (called the "atomic number"). The relative position of each element within the table can tell you something about how it may react with other elements (or in the case of noble gases, that they may not react much at all). However, these are only a listing of elements we know about, and in the exotic corners of our massive universe, many interesting things may be conceived.

It is possible that more elements exist that we have yet to discover. So the real question should be: what are the elements we haven't discovered?

The number of elements that could possibly exist is dictated by their relative stability. Right now, we know of 118 elements—twenty-four of which have been created in a lab. You see, the larger an atom becomes, the bulkier its nucleus is. As more and more protons pack into the nuclear space, their electromagnetic repulsion from one another makes it more difficult to keep the nucleus from bursting apart. So some of the heaviest elements that we currently know of had to be created in a lab because their nuclei are so unstable that they only exist for a short time before they begin to decay into other elements. This means that you aren't going to find them simply lying around Earth's surface. So we smash heavy elements together like balls of clay and see what larger elements we can create . . . even if only briefly (in some cases, fractions of a second).

Conditions may be extreme enough in certain areas of the universe to create the kind of power necessary to pound the configuration of larger atoms into place naturally. However, we can't hop in a craft and go on an atomic treasure hunt. So we are left to try to replicate the creation of those elements here. But the larger the element, the more power and creativity our scientists need to keep it together. This is something that is becoming more and more difficult as we reach the maximum threshold for maintaining synthetic element stability. So we need to either develop better technology to sample them from extreme conditions around the universe, or develop better technology to stabilize them on our own planet. In either case, we're limited by the infancy of our current innovation. Which is probably the case for *all* scientific quandaries we're trying to solve.

HOW DO CELL PHONES WORK TO TRANSMIT CALLS?

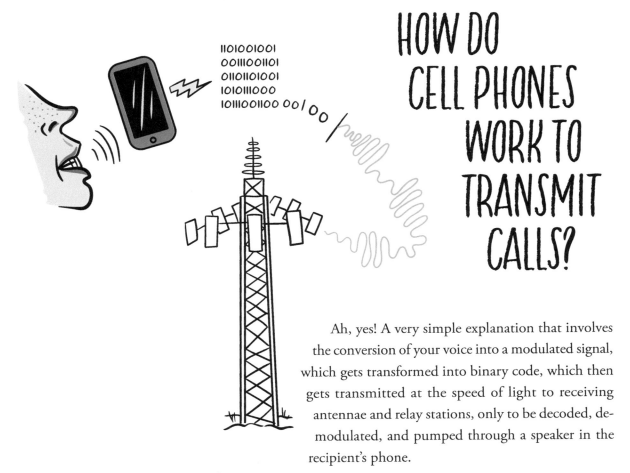

Ah, yes! A very simple explanation that involves the conversion of your voice into a modulated signal, which gets transformed into binary code, which then gets transmitted at the speed of light to receiving antennae and relay stations, only to be decoded, demodulated, and pumped through a speaker in the recipient's phone.

Simple, right? Woof. Okay, here we go . . .

For digital audio, the sound of your voice needs to be converted into a usable electronic signal. When you speak, the sound waves from the different pitches and rhythms of your voice move magnetic components within the microphone of your cellular device. The movement of these components is associated with a pattern of specific voltage changes that correspond to each nuance of the sounds received. Your phone then converts these logged voltage changes into binary code—a massive series of ones and zeros that represent the electronic blueprints for the sounds in the sentences you just uttered. These ones and zeros are then transcribed into radio waves and sent off at the speed of light to a nearby cell phone tower.

"But, Leah—how do you send ones and zeros through the air?"

Well, the pattern of the ones and zeros corresponds to certain altered and unaltered radio waves. This alteration can include changes to the height of the radio wave, the frequency of the transmitted wave, or its phase (which you can think of as changing its directionality—if the waveform points upward, you would point it downward, and vice versa). The pattern of alteration, and the subsequent alternating of wave types, corresponds to the ones and zeros from the original electronic code.

These radio waves are then received by a local tower, which sends the signal to a separate base station, where the signal is then relayed to its destination (which could be a direct shot or involve a volleying of the signal through a few additional stations and towers—this really depends on where you are). The station local to your phone call recipient will bounce the radio waves to an antenna in their phone. These altered and unaltered radio waves will then get converted back to binary code, which is then interpreted as a pattern of changes in voltage, which then evokes corresponding digital noises from the recipient's speaker (so basically the reverse process of what happened on your end during the coding phase).

Quite a series of events, and don't forget, they all happen in real time. The even wilder part? Once the sound waves from the device's speaker enter the ear of the recipient, there is a mechanical movement of modular parts within the ear that corresponds to a particular electric signal, which is then shuttled to the brain, where it is subsequently interpreted as the voice of the caller.

It's instances like this that make it hard to deny that the universe, and everyone in it, is just a part of one gigantic computer program. You know what I mean?

IS TELEPORTATION POSSIBLE?

Teleportation, in a sense, *is* possible and scientists have already begun fiddling with it. But it likely looks far different from what you're envisioning.

If I were to ask you to describe teleportation, you might depict *Star Trek*—a complete molecular disassembly at some origin location, with a subsequent reassembly somewhere else (like back on the starship *Enterprise*, for instance). Or perhaps the idea of being picked apart by an energy beam makes you uneasy (with good reason). So your version of teleportation looks more like walking through a doorway, intact, with emergence from another doorway at a distant location.

In either case, you'd be way off.

The kind of teleportation that scientists are currently flirting with involves something called "quantum entanglement," and it transports *information* rather than matter. Now, I warn you, it's a very bizarre premise, but one that has been observed in multiple studies,[11] using very small stuff (in fact, the Nobel Prize in physics was just awarded for work in this field). Quantum entanglement essentially links things together (as currently published, we've linked photons, calcium ions, electrons, and even bacteria), so that the state of one tells you something about the state of the other(s).

Let's frame it simply like this: I've got two plastic cups and two marbles—one marble is black, and one marble is white. I tell you to turn around as I hide one marble underneath each cup. When you turn back around, I ask you to reveal one of the marbles under only one cup. The marble is white. Without even looking under the second cup, you already know that the other marble is black. Seeing the state of one marble tells you what the state of the other marble is—their states are mutually exclusive.

Quantum entanglement is a bit like this, except our two marbles in this thought exercise are very special marbles—they act more like subatomic particles. So they are not simply a predetermined black or white—they are each, in fact,

DID IT WORK, GUYS?

PHYSICS 115

both black and white simultaneously (a property of something called "superposition"). It isn't until you lift the cup and observe one of the entangled marbles, allowing it to assume a specific color, that you send information to its partner and solidify its choice of color as well. So you see, in quantum entanglement, particles not only have mutually exclusive states, but their states are also intrinsically linked. They *cannot* exist in the same state at the same time. Regardless of any extraordinary distance that may separate them. So if I were to ask you to observe the state of a marble in my lab, and you tell me that it's white, this information is instantaneously shared with its now-black, entangled partner. Their exclusivity still works even if we were to have a lab assistant hide one of the entangled marbles hundreds or thousands of miles away—they maintain this strange linkage, share information instantly, and always "know" what their partner is doing. This is the weird world of quantum entanglement.

What does this have to do with teleportation? Well, by using this exceptionally powerful mode of linkage, we would be able to "teleport" information instantly, even across light years. The hope is to capitalize on this kind of instantaneous transfer to bolster the processing speed in quantum computing, turning the static ones and zeros of our current binary coding systems into something far more powerful. Why? Well, in effort to solve the really big problems (how to cure cancer, how to reverse climate change, how to efficiently transfer societies to clean energy) we need more robust computing.

The best part of quantum entanglement is that Albert Einstein, in all his glory, scoffed at the idea. It deeply bothered him. He disparaged it, calling it "spooky action at a distance." Paradoxically, he was also on the team of scientists who initially proposed the concept. Because what is the development of unprecedented scientific theory without a little intermittent hatred for it?

IF I FIRED A BULLET STRAIGHT INTO THE AIR, WOULD IT COME BACK DOWN AT THE SAME SPEED?

What goes up, must come down, right? Which, if we're being honest, is one of my least favorite adages. I have a feeling it was created in response to someone witnessing something painfully obvious too. I picture a man, who is the spitting image of Ignatius Reilly from *A Confederacy of Dunces*, mouth-breathing, watching someone climb a tree and subsequently fall out of it. "Hyuck! Well, gosh! What goes up must come down!" It's a little on the nose.

Anyway, back to the physics! A bullet from a standard 9mm handgun will leave the muzzle at a velocity of approximately 1,200 feet per second. If you were to fire the gun vertically, its velocity upward would immediately begin to decrease as it fights against the joint efforts of gravity and atmospheric friction. After a few thousand feet of propulsion into the air, these forces would slow the bullet to a momentary stop, and then gravity would drag it back down to Earth's surface.

When the bullet begins its descent, it is starting from that brief instance of no

FIRST TIME...?!

PHYSICS 117

velocity. So even though it was rocketed upward by the gun blast, it would still begin falling from the top of its arc at a state of complete standstill. The physics would be the same if the bullet had been dropped from someone standing at the top of a four-thousand-foot ladder. As the bullet descends, it will fall faster and faster, propelled at an acceleration rate of 9.8 m/s^2, which is the acceleration due to the force of gravity. But the faster this bullet falls, the harder the force of air resistance that opposes it. Eventually, the force of air resistance will balance with gravity, and the bullet can travel no faster on its way down—this is known as "terminal velocity," which is approximately 200 miles per hour (240 kilometers per hour) for our 9mm bullet.

So the answer to your question is no—even if you fire a bullet directly into the air or shoot a friend upward out of a cannon—if the projectile has enough time to fall, it will only reach the maximum speed of terminal velocity. Which, in the case of a 9mm from a standard handgun, is a fraction of the speed from the initial shot. I still don't recommend shooting directly into the air, though. You probably shouldn't launch your friend out of a cannon either.

IS IT TRUE THAT A FEATHER AND A BOWLING BALL WILL LAND AT THE SAME TIME IF YOU DROP THEM ON THE MOON?

There are moments of blinding poetry in science—discoveries that shift the tides of human history forever . . . And then there are moments when you spend millions of dollars to have a guy drop a hammer and a feather at the same time.

This is that story.

In 1971, Apollo 15 mission commander David Scott stepped out onto the lunar surface. After conducting a series of analyses (and likely gawking at the view), Commander Scott was video recorded dropping a geologic hammer and a falcon feather simultaneously. As the camera shows, and as Galileo predicted hundreds of years prior, they both hit the moon dirt together.

The grainy recording is actually on YouTube and is definitely worth the watch. Along with a hysterical compilation of astronauts falling while walking in their moon suits (thank me later).

This phenomenon is counterintuitive to us terrestrial children—on Earth's surface, feathers float on the breeze; hammers decidedly do not. But the fundamental difference between Earth and the moon? Earth has an atmosphere.

Our atmosphere contains a multitude of gases, vapors, and particulates: a sea of tiny airborne bits of matter. When you drop a feather on Earth, this matter-filled atmosphere pushes against the fine, branching fibers of the feather and opposes its descent. A bowling ball (or a hammer, in the case of Commander Scott's experiment) is dense and lacks the surface-area-to-mass ratio necessary to evoke the same air brake as the feather. So on Earth, the feather gingerly floats, while the bowling ball crashes down on your toes.

The moon boasts a completely different dynamic: it has virtually no atmosphere and is a great setting to demonstrate physics within the vacuum of space (provided you have a life-support system with you). Without an atmosphere to apply air resistance, the feather and the bowling ball are dropped and subsequently accelerate toward the moon's surface, directed by the exact same gravitational force. Same overall force = same speed downward = hitting the surface at the same time. As simple as $F=G[(m_1 \cdot m_2)/R^2]$.

As a burgeoning nine-year-old scientist, I too explored the implications of the force of gravity by jumping out of a tree. Unfortunately, just like the bowling ball, I managed to garner very little benefit from air resistance. Which culminated in a sharp thud and a badly mangled wrist. I learned two things that day: (1) gravity is inescapably powerful, and (2) pretending to be a Power Ranger is ill-advised if taking leaps of faith from twelve feet up.

WHAT IS THE DIFFERENCE BETWEEN DIRECT CURRENT (DC) AND ALTERNATING CURRENT (AC) POWER SOURCES?

There are so many instances, in various fields of science, in which the naming of phenomena isn't exactly intuitive. Luckily, physicists are pretty practical people, and for the most part, they do a good job of labeling things. Direct current (DC) and alternating current (AC) are both fantastic examples of this.

A DC power source provides a steady current that flows in one direction with a consistent voltage; an AC power source provides a current that changes directionality and exhibits variable voltage as it flows. The outlets in your home are tapped into AC power; your electronics that operate on batteries (or a cord with an AC adapter) are run on DC power.

Direct current was the first kid on the block when it came to the utilization of wires to supply electricity for municipal purposes. Both DC and AC power need a big boost in voltage when traveling long distances, for a variety of reasons. However, that high, transport-level

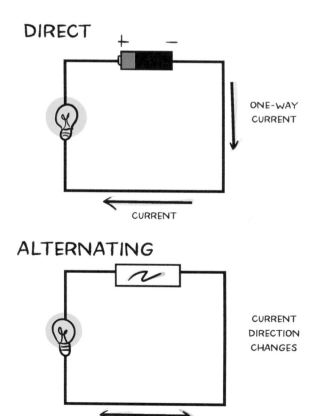

voltage is overkill when supplying a residence. So for both current types, the voltage needs to be decreased before routing it for civic home-to-home usage.

The properties of direct current make it tricky (read: expensive) to raise the voltage of direct current for transmission over long distances, only to subsequently drop that voltage before it gets pumped into a smaller local grid. In the past, making the widespread usage of DC power feasible required a hefty number of power stations—one positioned every couple of miles within a city grid, in fact. This made the utilization of DC power impractical, especially at the time. In response to this DC drawback, AC power was developed to make voltage conversion far easier and more cost-effective via transformers. After Nikola Tesla's AC patents were scooped up by a prominent entrepreneur, it was game over for the idea of a DC-based system.

Less sciencey, but unbelievably true history: Thomas Edison (American inventor; deep financial interest in DC power) and Nikola Tesla (electrical engineer; inventor of AC generation/transmission) had beef. Tesla actually worked for Edison briefly, helping him refine DC generator technology. During his employment, Tesla tried to sway his boss on the merits of AC power. Ultimately, Edison wasn't interested, and Tesla quit his job. Shortly thereafter, Tesla filed for multiple AC patents, which he then sold to a businessman named George Westinghouse. Cracking under the pressure of trying to prove the superiority of DC systems, Thomas Edison was reported to have allegedly electrocuted animals with AC power, in public demonstrations, in an effort to frighten communities and dissuade them from supporting the adoption of AC in their homes. Not his proudest moments, certainly.

HOW DO PHOTONS CARRY COLOR?

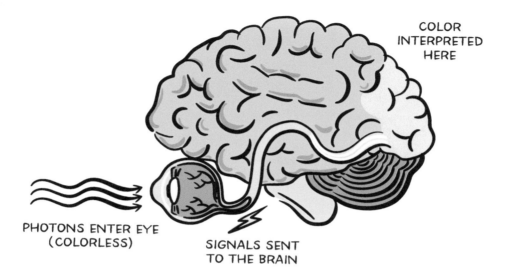

Color is a construct of neurological processing. The human eye is a sensory organ that transmits signals to the brain in response to exposure to a certain segment of the electromagnetic spectrum. So photons don't actually carry color—they are quantized units of radiation that we have evolved to be able to detect. It is our own minds that have assigned distinct hues to each of their separate wavelengths.

The rich colors we see every day are a product of how photons react with the environment around us. For instance, the standard number-two billiard ball appears bright blue because photons with a wavelength of approximately 450 nm are reflected from the pigments in the paint on the ball. Similarly, wavelengths of approximately 600 nm will bounce off the skin of an orange and appear to us to be a vibrant hue by the same name. These reflected wavelengths enter our eye and activate photoreceptors at the back of it, which promptly send signals into the brain for interpretation.

I suppose it's fortuitous that our brains have allowed us to interpret color from the wavelike properties of electromagnetic radiation. There's something slightly less flirtatious about telling your crush that their eyes reflect photons with the most lovely 550-nm wavelength.

HOW DO MICROWAVE OVENS HEAT FOOD?

If you have ever careened down a slide as a child in shorts, then you intimately understand the (painful) heat associated with molecular friction. In a similar capacity, microwave ovens work to heat food via the generation of friction on the tiniest scale.

The inside of a microwave oven contains a small device called a "magnetron." This device uses electricity to generate electromagnetic radiation with a frequency that corresponds to the wavelength of microwave radiation (which, to play the role of Captain Obvious, is how the appliance adopted its name).

In the cooking chamber of the appliance, the magnetron fires microwave energy through the food. As the electromagnetic radiation spins, twists, and otherwise vibrates the molecules within your pizza rolls, it increases their intermolecular friction (or how the molecules rub against one another). This physical work, applied by the microwaves, induces motion and releases heat. So in two minutes, you have a piping-hot plate of kinetically derived thermodynamics—yum!

Fun fact: the utility of using microwaves to heat food was discovered accidentally (as many great human discoveries are). A bright engineer named Percy Spencer was employed at a post–World War II radar plant. During one particular shift at the plant, he shoved a bar of chocolate into his pants pocket to be enjoyed later in the day. But while working near one of the radar tubes, he found that his tasty treat quickly melted into a tepid pocket-puddle. Thereafter, he went on to develop the very first microwave oven, which was marketed at the low price of $57,000 (by 2019 currency standards).

WHY ARE NUCLEAR WEAPONS SO POWERFUL?

"I know not with what weapons World War III will be fought, but World War IV will be fought with sticks and stones." This quote is attributed to Albert Einstein, who, while not a direct engineer for the atomic bomb, was a primary contributor to the equations that made it possible. He also understood the grave implications of this technological advancement in human armament.

The principles that underlie the devastation associated with atomic weaponry come from the famed equation $E = mc^2$. This equation tells us a few things about how our universe is set up: (1) mass and energy are different sides of the same coin, (2) mass and energy can be converted into one another, and (3) the conversion of a relatively minuscule amount of mass yields a tremendous amount of energy. The first two points probably sound a little strange to you; nevertheless, they are both facts of the universe. But try not to get hung up on *why* mass and energy are interchangeable—just know that mass harbors immense potential to make a loud boom. This point can be well demonstrated in the math of Einstein's equation.

$$E = mc^2$$

TOTAL ENERGY RELEASED BY BOMB = (MASS CONVERTED DURING FISSION REACTIONS) × (SPEED OF LIGHT [299,792,458 M/S])^2

Since I'm currently sitting on my couch, let's use my dog as an example. He is a healthy one-year-old pit bull of robust stature—currently, this young man is about 80 pounds (36 kg). If I were

PHYSICS 127

to convert *all* my pit bull into pure energy, we would use E = mc² to calculate a rough estimate of his energetic yield. It would look like this:

Total energy = [weight of my dog] x [speed of light]²

Plugging in the numbers:

Total energy = [36 kg] x [299,792,458 m/s]² = 3,235,518,643,452,543,504 J

Based on historical estimates and the fractional fission observed in warheads, this means that the mass of my dog would yield over two hundred thousand times more energy than the atomic bomb that was dropped on Nagasaki, Japan, in 1945.

You may now pick your jaw up off the floor.

Atomic bombs trigger a reaction called "nuclear fission," wherein the nucleus of an atom is split, and some of that lost nuclear mass yields energy. Upon detonation of a bomb, the radioactive isotopes (typically uranium or plutonium) undergo a rapid chain reaction. When one of these nuclei is split into lighter atoms, it releases energy and subatomic particles called "neutrons." These neutrons careen into other nuclei, splitting them and yielding more neutrons and even more energy. This fission process quickly becomes an unstoppable and highly destructive cascade of exponentially increasing energy release.

In the radioactive fuel of an atomic bomb, trillions and trillions of atoms release energy, more or less at once. This is why as little as 2.2 pounds (1 kg) of radioactive mass inside of a warhead (of which only a fraction undergoes fiery fission) can generate enough energy to obliterate everything within a one-mile radius.

ARE THERE MORE COLORS THAN ROYGBIV?

For those of you unfamiliar with the initialism, ROYGBIV lists the colors in our visible spectrum of light: red, orange, yellow, green, blue, indigo, and violet. The photoreceptors in our eyes are responsible for our perception of these colors. Interestingly, these photoreceptors are also responsible for our *lack* of perception of other potential colors as well.

The light that we *can* see constitutes a relatively narrow range of wavelengths within the electromagnetic spectrum. This electromagnetic spectrum describes the full range of light radiation in the universe (as far as we understand). The basic unit of this radiation is the photon. While you may already be familiar with photons, what you may not know is that they convey a wide range of energy levels. In fact, photons aren't just the stewards of our visible light, but they also comprise long, low-energy radio waves, all the way to super short, penetrating gamma rays.

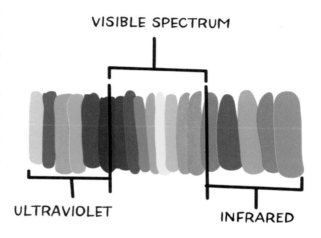

Our eyes have evolved to detect electromagnetic radiation with wavelengths between 740 nm and 380 nm. Within this narrow range, certain smaller wavelength strata are interpreted by our brains as separate colors. But there's nothing special about the range we detect—had we developed additional photoreceptors, we may have been able to detect ultraviolet, infrared, or maybe even X-ray photons as well!

All in all, the light range we see only accounts for approximately 0.0035 percent of the total electromagnetic spectrum on a linear scale. As such, we are bathed in a sea of photons to which we are almost completely blind. Makes you wonder what you might be missing, eh?

COULD YOU FLY AN AIRPLANE UP INTO SPACE?

You could certainly try to fly a standard commercial airplane into space. But physics would work in opposition to this stunt, and I would be inclined to advise you against it.

Lift is an essential physical principle that directs a clunky large plane to lurch upward into the sky. Commercial airplanes are designed to ensure that air will flow faster over the top of the wings compared with the airflow beneath the wings. This creates a pressure differential between the top and bottom, which results in a net upward force. With enough speed, the force of upward pressure will overshoot the force of gravity, and then, dear reader, we have liftoff!

The pressure that creates lift is a product of the shuffling of disturbed molecules within our atmosphere. For lift to work, per the engineering of commercial airplanes, these combined molecular forces are essential. In the vacuum of space, or even the upper layers of the atmosphere where the air is incredibly thin, lift cannot be evoked in the same way. So if the aircraft were somehow able to make it that far above Earth, it wouldn't be able to maintain its high altitude.

In addition to lift, commercial airplanes also rely on propulsive force to fly. Inside a jet engine, the air is mixed with jet fuel to create combustion. The resultant hot, expanding gas is blown out of the back of the engine by turbines, which results in forward thrust. The air intake is critical to make the combustion reaction happen. Again, in the vacuum of space or a region in which the air is too thin, the engine would not operate.

So if you *do* decide to try to fly your commercial aircraft into space, the aircraft would not produce lift or thrust. You'd be pretty efficient at falling, though!

PHYSICS 131

HUMAN PHYSIOLOGY

"LAUGHTER IS THE BEST MEDICINE—UNLESS YOU'RE DIABETIC,
THEN INSULIN COMES PRETTY HIGH ON THE LIST."
—JASPER CARROTT, ACTOR

WHY IS BLOOD RED?

Human blood is red for a reason similar to Martian soil being red, or for that chain on your old bike (the bike you kept swearing you were going to bring in out of the rain) also being rust red. In all these scenarios, the ruddy hue is linked to the oxidation of iron.

The almighty carrier of oxygen within the human body is the red blood cell. To accomplish its important task, the red blood cell must employ the help of a protein called "hemoglobin." Hemoglobin contains iron, which it leverages to better capture oxygen atoms (iron happens to be positively charged, which attracts the relative negative charge of oxygen). When oxygen is bound to the iron inside hemoglobin, the combined structure reflects red wavelengths of light. Overall, this makes blood—most especially oxygen-rich arterial blood—appear to be bright red.

I suppose this begs the next question: since other types of metals can also be oxidized, can there be other colors of blood? Fantastic question, and absolutely yes! For example, octopuses have evolutionarily elected to utilize a protein called "hemocyanin" to carry their oxygen. Instead of iron, these proteins contain copper, which helps the cephalopod transport oxygen in cold, oxygen-poor environments. By similar principles, the copper atoms will also become oxidized. But, when copper links to oxygen inside of hemocyanin, the structure reflects bluish-green wavelengths of light, and by extension, this contributes to octopus blood appearing blue.

I was a physically reckless kid. I climbed trees, careened through treacherous terrain on my skateboard, and used old PVC pipe sections like makeshift bazookas to try to shoot my friends with pop bottle rockets—standard stuff. What all of this means is I have sustained a few nasty injuries throughout my life, and my blood was always red. But I remember being told by my friends that the blood inside of veins is blue. There was even one kid in my neighborhood who swore he had seen blue blood when his little brother was bitten by a dog. Well, I'm here today to tell you that this is false—you're a liar, Justin. You lied.

BASICALLY RUSTY IRON

HUMAN PHYSIOLOGY 135

WHY IS THE HAIR ON MY ARMS SO SHORT, WHILE THE HAIR ON MY HEAD KEEPS GROWING?

It's you and me, dear reader. We're answering life's big questions.

All cells in the human body are *not* created equal. Especially when it comes to life span. Cells actually have individual biological timelines that hard-code a death date into the DNA they contain. Like super macabre clockwork, the death date is inescapable (unless you're a cancer cell, which has mutated to exhibit a practically infinite life span. In which case: fuck you).

Cell life spans are highly variable and highly important, and they are a product of their function within the body. For instance, cells that line your gastrointestinal tract have an incredibly short life span of three to five days. Which is a good thing! This high turnover rate ensures that the caustic slurry of digestive juices that pour through your gut-pipes don't eat through that cellular barrier. Fresh cells = freshly reinforced guts and a decreased risk of ulcers. Osteocytes (bone cells), on the other hand . . . or foot . . . or leg . . . (this is a poor attempt at a joke about bones), live for up to twenty-five years. Which again is fortuitous—you would likely not fare well walking around on transient bone stock.

This now brings us to the flowing Pantene Pro-V locks on your head and the associated little wispies that cover the rest of your body: why the disparity? Well, at some point in our evolutionary past, our (incredibly) hairy ancestors climbed out of trees and began walking upright. This bipedal method of movement brought about a whole host of changes for our distant family: larger spans of territory, new diets, and far less body hair.

But why the nakedness? While this topic is still being discussed among evolutionary biologists, the most widely accepted theory is we simply got too hot! As we moved from our cool, shady homes in the treetops, we began scampering across grassy plains in search of food. Under the blazing sun, the extra hair was too toasty. So through the course of generations, those of us with more sweat glands and thinner hair were better at thermoregulation (controlling body temperature). Fast-forward a few million years, and here we are: the most shameless among the great apes.

"Yes, but Leah: why did the hair stay on our noggins?"

Understandable question. While certain adaptations are obvious (gills for breathing underwater, for instance), others are a bit more speculative. As far as head hair is concerned, the consensus is that it is used for insulation and protection of our sensitive scalps from UV radiation.

To answer the original question, though: these two types of hair cells have evolved to maintain drastically different life spans. Body hair terminates growth after a couple of months (and thus stays short); head hair terminates growth after several years (and can be subsequently styled into a most majestic mullet).

WHAT IS THE PURPOSE OF PUBIC HAIR?

It turns out that having a 1970s-era, poofy hairstyle between your legs may provide a *bush*el of advantages.

First, the hair confers a protective barrier from exogenous microbes, dirt, and debris. Similar to the role of eyebrows and eyelashes for your eyes, pubic hair can help filter out particulates before they reach the delicate genital tissues. This helps keep the area cleaner and reduces unwanted mechanical abrasions. Also with respect to microbes, oils produced in the genital region, maintained by a healthy covering of hair, may help inhibit unwanted bacterial populations. In a study published in 2021 by Geynisman-Tan, et al., it was even found that going from "hair-to-bare" changed the composition of healthy vaginal bacterial populations in a group of sixteen women when compared to a control group.[12]

Second, pubic hair helps reduce friction during day-to-day activities and . . . well . . . "nighttime activities" as well. The fragile tissues of the genitalia can become irritated and chafed from kinetic friction associated with sexual intercourse. However, it has been shown that the coefficient of friction at a hair-to-hair interface is smaller than that with direct skin-to-skin contact. So this makes pubic hair in both partners a sort of nonwet lubricant.

Third, the development of public hair is a visual cue for sexual maturity. In our ancestral past, this may have served as a confirmation of the prospective readiness of a partner for viable mating. This signal could have been an important cue to guide mating in a younger population of humans, increasing the likelihood of partner fertility and possibly leading to more successful reproductive practices to keep our species going.

So while pubic hair removal is quite common in the modern era, just know that your nether-fluff works hard to keep your most precious parts as healthy as possible.

WHY DOES ASPARAGUS MAKE MY PEE SMELL BAD?

The most interesting thing about this entry is that many people are going to read the question and say, "Wait—asparagus makes your pee smell bad?" But let's dive into this biologically based incredulity a bit later. First, let's talk about the biochemistry!

There is a unique compound found inside asparagus, aptly named "asparagusic acid," that contains a couple of sulfur atoms locked inside of a ring structure. When asparagus is eaten, some of the bonds within the asparagusic acid are broken down, the compound is chemically altered, and a few different sulfur-containing byproducts are produced. These compounds get excreted in the urine, and they quickly vaporize into the air, eliciting a pungent, almost rotting odor. Sulfur compounds are pretty notorious for this—scents like skunk, onion, eggs, and flatulence are just a few terrible examples attributed to these molecules as well.

It used to be assumed that the reason some people are ignorant of the post-asparagus-binge urine stink is that they didn't have the digestive machinery to break down asparagusic acid. However, evidence is mounting that suggests it's actually a gene mutation that prevents people from detecting the odor altogether. So it's not necessarily that the smell doesn't emanate from them, but only that they don't produce the receptors in their olfactory system to be able to pick it up.

So if you're going to have a genetic mutation, I think that not being able to smell the sharply putrid scent of fresh asparagus pee is probably a good one to request.

WHAT CAUSES STUTTERING?

Stuttering is characterized by a disruption to the normal flow of speech patterns. This typically manifests as the repeating of a portion of a word, repeating of the whole word, or frequent pauses that obstruct the cadence of speech. In young children, this can be a normal part of learning how to speak as sensorimotor connections are still being strengthened. But in adults, it is quite abnormal.

While most of us don't give the mechanism of speech a second thought, the physical process is quite complex. It is an intricate dance between neurocognitive processes, motor control of a multitude of anatomical structures that make speech happen, and sensory feedback (primarily auditory) that helps the brain correct or adjust speech patterns on the fly. So speech is a simultaneous execution of motor signals leaving the brain and sensory signals coming in. All of which need to be managed and processed in parallel for speech to take place appropriately.

Due to the complexity of these speech mechanisms, it is likely no surprise that stuttering can be the result of any number of breakdowns within the sensorimotor chain. For this reason, precise underlying causes for this disorder are still being explored. Some evidence suggests that many cases of adult stuttering may begin with an impairment to the system of auditory feedback. If feedback received by the brain is delayed or distorted, it can severely inhibit fluid downstream control of the motor part of speech, resulting in repetitive articulation.

The utility of being able to self-regulate speech cadence via that auditory monitoring cannot be underestimated. Challenge yourself: Try to communicate something to someone else from across the room. Then, without breaking cadence, continue to communicate with them while having a second person talk to you through a pair of loud headphones. I'd be willing to bet that scrambling your auditory feedback would lead to speech pattern disruption. And possible confusion from one or both people you're attempting to communicate with.

WHY DOES MY EYE RANDOMLY TWITCH?

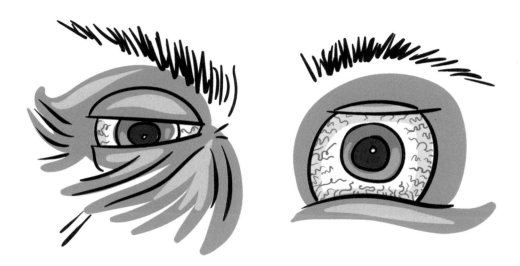

While there may be multiple reasons for twitching eyelids, today we will talk about the one with a benign etiology.

The annoying fluttering sensation that you may occasionally feel around your eyes has a fancy name (of course): eyelid myokymia. From a physiologic perspective, the affected muscle group around the eye experiences uncontrolled electrical discharges from the fine nerves that supply it.

Under normal circumstances, when a muscle group is activated by its associated nerve(s), it contracts, pulling bones or soft tissues into place. These contractions are controlled and elicited when contraction is required for necessary movement. With eyelid myokymia, the contractions typically occur in variable frequency outside of normal functionality and can be quite disruptive to the affected individual.

While the underlying causes of benign eyelid myokymia are poorly understood, studies have indicated that its occurrence is associated with lack of sleep, excessive caffeine intake, and increased emotional stress or anxiety[13]—in other words, the primary risk factor is being an author.

WHY DO I SEE RANDOM COLORS WHEN I PUSH ON MY EYES?

I know what you're thinking. You're wondering, *Okay, who submitted this* . . . But I also know that you know exactly what this adult human is talking about. And now you realize you're *also* delightfully weird (welcome to the club).

This phenomenon of rubbing your eyes only to see a kaleidoscope of colors is actually a rapid firing of sensory receptors at the back of your eyeball. The shifting shapes and neon hues you perceive are called "phosphenes."

The human eye is an organ absolutely jam-packed with specialized cells whose only function is to collect light, become activated, and subsequently send electric signals to the brain through the visual superhighway of information (the optic nerve). The receptor cells are located in the very back of the eyeball in a spongy, pink area called the "retina." When light enters the eye through the pupil, it lands on the retina and activates the receptor cells embedded within it. Once activated, these cells communicate with the brain through various

HUMAN PHYSIOLOGY 147

transfers of electricity between neurons. Much like a computer, the brain interprets the patterns in these electric signals and forms the image in your mind of what is being viewed.

So if your eyes are closed and the retina isn't being exposed to light, why are phosphenes even a physiological thing? Well, the receptors in your eye are incredibly sensitive (which is what allows us to see vague shapes even in the dark). They are so sensitive, in fact, that even minute changes in pressure can set them off. So as you rub your sleepy eyes, or hard-blink incredulously at the email you received from a coworker, you are increasing the pressure within the eyeball. This, in turn, induces unexpected, additional pressure on the receptor cells, and it sets them off. However, in this capacity they aren't being activated by light, so the signals being sent to the brain don't make enough sense for the brain to form an image of anything specific. The best it can interpret is a smattering of amorphous shapes and flashes of color.

Science of phosphenes: definitely something to bring up during a first date. (Disclaimer: may not guarantee a second date.)

WHAT IS CANCER?

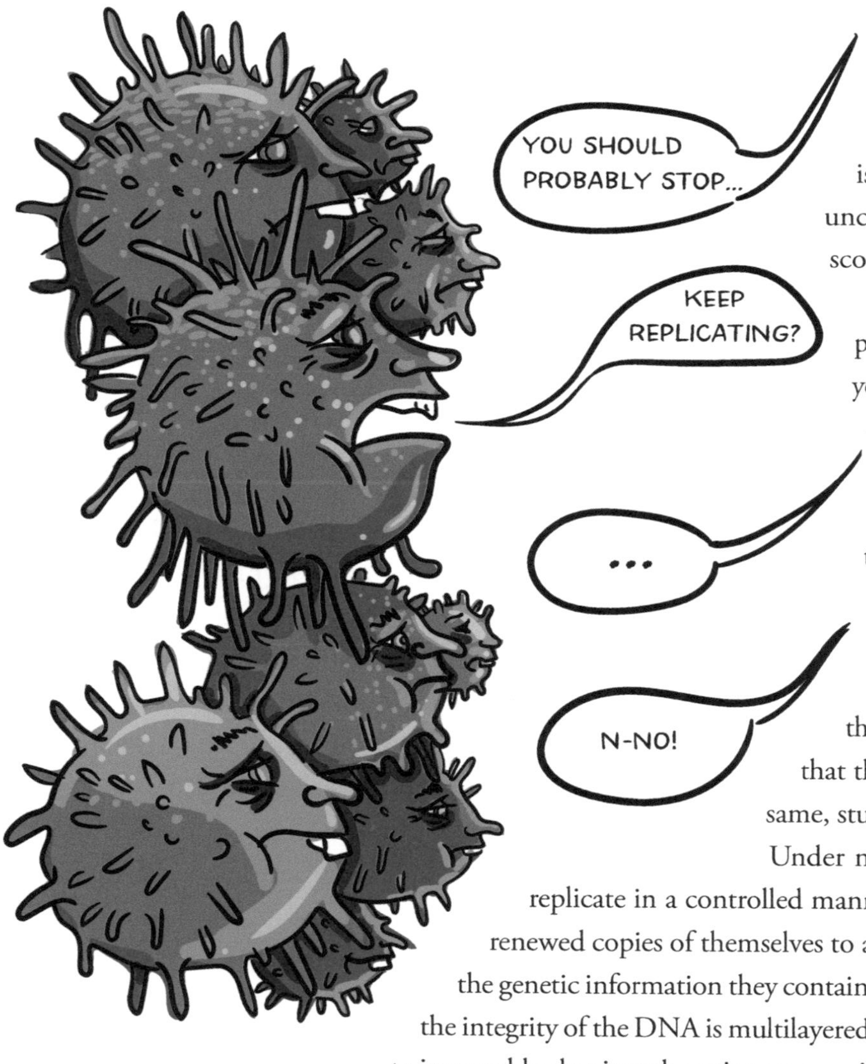

"YOU SHOULD PROBABLY STOP..."

"KEEP REPLICATING?"

"..."

"N-NO!"

As a former researcher in the field of clinical oncology, I feel like I'm qualified to say this: cancer is a dickhead. A punishing, uncaring, unwanted biological scourge of a dickhead.

Now, from a scientific perspective, cancer is part of you. At its base origin, it is a cell. A cell that demonstrates a massive failure of the elegant checks and balances that are supposed to keep your cells from becoming cancerous. Basically, cancer is a cell that has been left on the biological photocopier and that then fills your body with the same, stupid image of itself.

Under normal circumstances, cells replicate in a controlled manner. As they age, they make renewed copies of themselves to avoid deleterious damage to the genetic information they contain. Now, the system to protect the integrity of the DNA is multilayered and incredibly complex. So to irrevocably alter it to the point cancer arises, multiple cellular security guards need to be forcefully taken offline. This mechanism of aberrant behavior is called the "multihit model," and it describes the nature of how cancer evades multiple layers of preestablished checkpoints.

While the pathophysiology and genomic cavalcade of cancer are complex and highly unique to the individual human and type of cancer expressed, all cancers follow—more or less—the same dance steps (in no particular order):

STEP 1: Locate the nearest tumor suppressor gene (of which there are many) and promptly put it to sleep.

STEP 2: Make good with the local genes that stimulate cellular proliferation and growth; encourage these blokes to kick their activities into hyperdrive.

STEP 3: Rinse and repeat.

After enough of the watchdogs have been effectively silenced, the cell begins to replicate. Over and over and over and over again. Without anyone telling it to stop. This cellular snowball effect leads to tumor formation.

While I certainly don't have enough room to cover the mechanisms of genetic damage responsible for cancer, I do want to tell you about something you've likely seen mentioned on cigarette boxes: carcinogens.

(Side note: The tobacco industry purposefully uses lengthy words to confuse the consumer. Carcinogen means "something that causes cancer." But the PR teams at Marlboro didn't think that was the sexiest description to put on a product they were attempting to sell for a profit.)

Carcinogens are chemicals known to alter the genome in a way that increases an individual's risk of cancer. The reason that something like smoking or being exposed to caustic fumes from processed fossil fuels is so dangerous is that they are packed to the absolute gills with carcinogens. And based on the multihit model I described, the more exposure an individual has to things that can mutate DNA, the more pieces of the protective puzzle get chipped away. It boils down to a numbers game and an effective risk: the higher your DNA mutation count, the more likely it is that you happen to bear the right sequence of unfortunate genetic blips to allow for unchecked cell copying.

Like so many of our human diseases, the ultimate cure for cancer will likely come down to genomic editing and correcting the mutations at their source. Cutting-edge medical stuff. And also a great piece of sciencey context if you want to make a zombie-outbreak movie.

Get at me, Hollywood.

WHICH ORGANS CAN HUMANS LIVE WITHOUT?

The slightly disturbing nature of this question aside, it's an interesting thing to consider. There are many kinds of organs (for instance, your skin happens to be the largest organ of the human body), but I think you may be curious about the viscera locked away inside of your body cavities—let's cover those. The best way to answer this is to separate these organs into groups: those that come in a pair and those that do not.

Organs that come in pairs may continue to function without their partner. While missing one organ in a set may adversely affect the physiologic efficiency and workload of the remaining organ, life can still be sustained. These organs include lungs, ovaries/testes, and kidneys.

There are many more organs that are naturally unpartnered. However, the human body can still function in their total absence. These organs include the spleen, stomach, bladder (which can be rerouted in a couple of different ways), combined structure of reproductive organs, gallbladder, appendix, colon, and thyroid gland (with supplementary pharmaceuticals).

The liver is a special case because, while you cannot survive without a liver, you can still live if a substantial proportion of the liver is removed. Studies from members of the transplant team at the University of Pittsburgh have estimated that normal functionality of the liver can be maintained with as little as 25 percent of the organ intact.[14] The most fascinating part is that barring irrevocable damage to the original tissue, the liver can regenerate back to full size.

The relatively safe removal of many of these vital body parts is made possible with modern technological advances. So while their removal is livable, this may not necessarily be the case without proper medical oversight. I suppose I saw fit to make that disclaimer in the event you felt like removing any of your organs.

WHAT IS A HANGOVER?

A hangover is a fairly strong indication that when you explained to your friends you were "totally good" to take another shot of tequila, you were, in fact, *not* totally good.

Hangovers are a cascade of physiologic events that, when combined, make for a rough day following a bender. Induced by overconsumption of alcohol, the common hangover includes symptoms that run the physical and mental gamut, from headaches and dizziness and muscle aches, to fatigue and clouded cognition, to nausea and vomiting, to possible abnormal blood pressure and heart rate. Alcohol contributes to these unpleasantries in several ways.

Alcohol consumption can lead to substantial dehydration. Not only does alcohol drastically increase the renal output of urine, but it is also associated with gastrointestinal upset (like diarrhea or vomiting). This loss of fluid via compounded pathways unfavorably impacts water and electrolyte balance.

Alcohol can contribute to elevated blood sugar. While this is typically observed in more chronic drinking states, alcohol consumption may contribute to poor nutrition intake and may also fundamentally alter metabolic processes for several organ systems. For instance, chronic alcohol usage contributes to decreased functionality of the liver and a subsequent uptick in the buildup of lactic acid and alters the body's ability to efficiently utilize available glucose.

Alcohol negatively impacts slumber. This may seem counterintuitive because alcohol is a well-known central nervous system depressant. So while the onset of sleep may be easier when inebriated, the *quality* of your sleep patterns becomes disrupted. Following alcohol consumption, the REM sleep cycles (periods in which you typically dream) are shortened, with corresponding extended periods of deep sleep cycling. This leads to an imbalance of normal sleep patterns. From a behavioral perspective, alcohol consumption and social drinking typically also happen at night, competing directly with normal sleep hours. Combined, these factors help explain why you typically feel exhausted the morning after you party with your closest cohorts.

I don't typically drink alcohol. I may have a glass of red wine on my birthday, which I will sip for the duration of dinner, but that's the extent of things. When people ask me why I don't drink, I think they expect me to express a self-entitled, elevated sense of philosophic being. But really? I just don't want to feel that throbbing headache-hammer of regret ricocheting around inside of my skull for twenty-four hours. I've done that already. Many, *many* times. Besides, in my youth, social media wasn't really utilized to the extent it is today. So I could let loose and not have to worry about being documented. These days, I'm worried about being photographed clumsily performing the "Cha Cha Slide" on bar furniture to the grave chagrin of any unfortunate onlookers. I guess not much has changed for me in the last fifteen years.

HOW DOES PAIN MEDICINE (ACETAMINOPHEN) WORK?

During my graduate pharmacology and biochemistry courses, I was always darkly amused by the frequent caveats for drugs printed in textbooks: "mechanism of action remains unclear" or "mechanism of action currently being explored." You would think these lines would be included for investigational pharmaceuticals only, right? But you might be surprised (or appalled) to know how many commonplace drugs also include these sneaky asides. Acetaminophen is one of those drugs.

Acetaminophen is marketed as a drugstore pain reliever and fever reducer (or in fancier language, an "analgesic and antipyretic"). While the specifics of its tricks and actions on the body are pretty muddy, we think that it affects something called cyclooxygenase (COX) pathways. These pathways manufacture lipid molecules (called "prostaglandins") that are partially responsible for mediating pain triggered at sites of inflammation in the human body.

It is believed that, at some point in the production line for prostaglandins, acetaminophen throws a wrench in the biological manufacturing works. This inhibits part of the COX pathway, which in turn significantly reduces the number of available prostaglandin molecules—decreased prostaglandins lead to decreased pain. Acetaminophen is metabolized by the liver and only has a half-life of approximately five hours. Once its availability dwindles, the prostaglandin production may bounce back to its expected output, and the pain returns. This is why acetaminophen only works as temporary pain relief.

While the description I gave is highly generalized, it also reflects the limits to our current extent of knowledge of how this pharmaceutical works. Since its patent in 1951, its usage is based on a "we don't know exactly how it works, it just does" strategy. And it is far from the only pharmaceutical immersed in this type of dubiousness.

WHAT IS PUS?

Pus is a bit like the remaining liquid carnage from a microscopic immunological battle. It is a thick fluid mixture of cellular debris (from microbial and host cells), odd-ended proteins, and expired white blood cells that fought valiantly against invading microbes. While most pus is a creamy off-white color, it can occasionally exhibit a tinge of green, brown, or red, depending on the microbes implicated in the local infection.

When the body detects a breach of miniature invaders, it immediately launches its army. A litany of defensive players is called upon to help mitigate the spread of infection. This includes macrophages (cells that eat foreign debris and microbes) and a few varieties of white blood cells. These cells release chemical signals to upregulate inflammation and recruit more immune cells to the battle site, helping to tip the scales against the infectious agents.

Most people attribute pus to bacterial infections, but it can also be produced in response to other unwanted invaders like parasites, fungi, or even viruses. While pus does not serve a functional purpose, it *is* a reliable visual indicator of infection. It also demonstrates the aggression of the host defenses—previously invisible immune cells becoming collectively visible due to their concentrated numbers, like the immune system's goopy, ivory-colored war medal.

Prior to understanding the importance of antiseptic practices (the mid to late 1800s), physicians and surgeons used to think that the presence of a certain kind of pus was a positive clinical sign. After they performed surgery—usually with their bare hands—they would actively look for white odorless pus a few days later, which they heralded as part of a healthy healing process.

Imagine consulting with a surgeon who, upon seeing your oozing surgical wounds, tries to convince you that your sign of active infection is a celebration-worthy finding.

MY WIFE'S BLOOD TYPE IS O, BUT NEITHER OF HER PARENTS HAS TYPE O BLOOD. SHOULD SHE BE CURIOUS?

I love this question because the submitter utilized my platform as a means to possibly initiate drama over the next family dinner. Unfortunately, dear reader, there will likely be no explosive scene or dangerously flung forks.

PEOPLE WITH A AND B BLOOD TYPES HAVE RED BLOOD CELLS WITH SURFACE ANTIGENS

O BLOOD TYPE DOES NOT HAVE SURFACE ANTIGENS

Blood type is determined by two genes—one is contributed to you by your mother, the other by your father. These gene codes form your ABO blood type. In other words, the coding combination of these two genes will determine if you have type A, type B, type AB, or type O blood. Important note: A and B genes always express themselves dominantly over O genes.

Because your mother and father each have two genes that determine their blood type, the one gene copy they each donate to you is chosen at random. Different combinations of these inherited genes will yield the following blood types:

BLOOD TYPE	GENE COMBINATION
A	AA
A	AO
B	BB
B	BO
AB	AB
O	OO

As you can see in the table, A and B genes overshadow O genes. So even if your genetic combination is AO, you will still have type A blood; if you have BO (the genetic combination, not the stink), you will still have type B blood.

In reference to the submitter's wife, her parents could have both been AO or BO, or perhaps one was AO and one was BO. Even if neither of them expressed type O blood, through random genetic combination, they both could have donated an O gene to their daughter, making her genetic combination OO and consequently giving her type O blood.

These factors being considered, I don't think the submitter's wife should be suspicious of the blood type discrepancy. But that shouldn't stop you from being problematic at dinner. I happen to love when people ask controversial questions at family gatherings!

WHY DO WE AGE?

The way this question is phrased elicits two possible responses. The first is to explain the biological *purpose* of aging; the second is the explain the biological *processes* that contribute to aging. I'm game to take a crack at both.

The reason our lives are finite is still being debated—I'm not exaggerating when I say that there have been hundreds of hypotheses meant to provide a rationale for our mortality. One of these hypotheses includes the facilitation of more robust evolutionary pacing. By having a limited life span, older members of our species quickly make way for new members. This ensures a trickle of fresh genetic variation with each generation, which could provide long-term, species-wide benefits such as better adaptability to changes in our climate, changes in food supplies, altered resource availability, resistance to disease, etc.

Another way to explain our expiration date is that the extent of our life coincides with how much damage our bodies can reasonably sustain before the cumulative detriment gets out of hand. Pollutants in the water, particulates in the air, ionizing radiation from the sun, and day-to-day cellular processes all contribute to molecular wear and tear on the human body. Decades of this mounting damage can lead to broken functionality of cells and proteins. This poorly functioning biology contributes to a myriad of chronic conditions and cancers. So our capped years help ensure that, even at the end of our lives, our bodies are still relatively functional.

Obviously, we're still trying to reason through why we've been engineered to shuffle off this mortal coil. But we're beginning to unpack the mechanisms for how this happens. One of the primary contributing factors of aging comes down to the culmination of genetic damage, with subsequent loss of healthy molecular-level functionality of proteins and cells. The DNA inside of your cells is a sacred blueprint that codes for every protein that carries out every function in your body. If you accidentally alter the instructions within these blueprints, it becomes damn difficult not to construct somewhat faulty molecules. So protecting the DNA is critical.

There are droves of support proteins and complex molecules whose sole purpose is to maintain the integrity of your DNA. One of these helpers is a noncoding portion of DNA that occupies space at the ends of your chromosomes. These are called "telomeres," and they are a bit like the plastic caps that wrap the ends of your shoelaces—they aren't a functional part of the shoelace itself, but they do help to keep the ends of the shoelace from becoming damaged and fraying. In a similar way, telomeres protect the ends of your DNA from sustaining damage, they act as an elongation buffer to keep the chromosomes from being shortened during the process of cell division, and they keep the free ends of the chromosome from being glued together by errant repair mechanisms. It's no surprise that these little biological endcaps take quite a beating.

As we age, our telomeres shorten with sustained damage and successive rounds of DNA replication cycling. At some point, the telomere becomes too short, compromising the integrity of the DNA. This can signal the parent cell to become dormant or to die—occasionally these distress signals are missed, and the cell may even become cancerous. Regardless of cell fate, the length of the telomere is a kind of biological fuse timer, shortening as a means of counting down to cell death and discontinuation. As our bodies age, more and more of these cells power down (something called "senescence"), global physiologic processes become less efficient, and we lose enough of our normal functionality that life is no longer sustainable.

Okay, after typing the cellular details of death, I feel like I need a pick-me-up. Videos of puppies wearing bowties to the rescue!

HOW DOES THE HUMAN SENSE OF SMELL WORK?

It may come as no surprise—especially if you've ever ridden in a fully packed subway car in the middle of summer—that the human sense of smell is incredibly acute. But I'm willing to bet that at some point in your life, you may have heard otherwise.

In 1879, an impactful French neuroanatomist, Paul Broca, published research in which he classified organisms into one of two categories: the "osmatique" (good sniffers) and the "anosmatique" (not good sniffers). Based on comparative anatomy alone, humans were lumped into the latter category. With some unsubstantiated support from famous friends like Sigmund Freud and others, the findings of this work spread like wildfire. This research forms the basis for the wild misconception that humans have poor olfactory senses, which persists today! Human beings may not have as many olfactory neurons or as much cortical brain tissue dedicated to scent as, say, a dog. But to suggest that our sense of smell is *weak* is inaccurate.

In the back of the human nose is a section of delicate tissue that houses as many as twenty *million* combined sensory neurons dedicated to scent. When you walk through something pungent—like freshly

HUMAN PHYSIOLOGY 163

cut grass—chemicals waft into your nose from the surrounding air. These chemicals dock into open receptors on the sensory neurons, triggering electrical signals to be sent to a little piece of the brain called the "olfactory bulb" (there are two of them—one for each nasal passage—and they look a bit like partially squished jelly beans). The olfactory bulb acts like a relay station: it receives electrical signals from the nose and then sends these signals to specific areas in the cortical brain for processing. The cortical brain is where these signals are interpreted. They may be mixed with signals from taste neurons to identify what is being consumed, or the signals may trigger a memory, a pleasant feeling, hunger, or (if we want to bring up the crowded subway car again) disgust.

The remarkable thing about our olfactory system? Recent estimates indicate that human beings can detect approximately one trillion or more distinct scents! So the next time someone says that the human sense of smell is pitiful, tell them about the millions of neurons and innumerable integrated neural pathways. Or just punch them directly in the face.

WHY DO WE EXHALE CO_2 IF WE INHALE O_2?

A few times throughout my stint as a science communicator, I have been asked why we breathe at all. Which is actually a great question. The act of breathing is pretty weird when you sit and think about it—rhythmic cycles of engulfing and expelling the atmosphere around you so that you can intermittently fill two pink bags that dangle inside of your chest. Strange.

The primary reason that we breathe is so we can more efficiently make energy. Oxygen (O_2) and carbon dioxide (CO_2) are both associated with a process called "cellular respiration," during which glucose is oxidized to make your all-important cellular battery (a molecule called ATP). When needed, the bonds inside of ATP are broken, unlocking the energy within them. This liberated energy is then harnessed to help drive innumerable critical reactions and mechanical processes throughout the human body.

The process of making ATP in the presence of oxygen increases the efficiency of production. Oxygen gets consumed during this complex cycle, and carbon dioxide is a by-product. So when we inhale oxygen, it gets shuttled from our lungs to our tissues, where it is utilized in ATP production. During the production process, carbon dioxide is released as a waste product, which is then transported in the blood back to the lungs and subsequently exhaled.

So our breathing cycles are a fancy way of importing oxygen to our cells so they can make energy, then exporting carbon dioxide, which is one of the waste products of that process. Like one big conveyor belt. That helps to make energy. By using your lungs as a kind of bellows.

Still a strange concept, even when the rationale is explained.

WHAT IS THE FUNCTION OF THE APPENDIX?

The appendix typically goes unnoticed by its owner, unless they start having completely unrelated, lower-right-sided abdominal pain. In that case, said owner begins to panic and attempts to calculate how many hours they have remaining before the appendix goes off like a subintestinal grenade. Most often, this fear gives way to the embarrassing realization that the pain is due, instead, to food truck–related gas. I'm sure we've all been there.

The appendix looks like a small deflated balloon that dangles haphazardly off the colon, near the area where the large and small intestines are coapted. It's pretty unremarkable to look at, as far as human anatomy goes, and it doesn't serve any immediate purposes critical to your core requirements for living. However, that doesn't make this teensy pink pouch completely useless.

It used to be assumed that the appendix was a vestigial part, something that may have been useful to us in our ancestral past but that has evolved to be defunct. However, evidence from recent research suggests otherwise. One theory indicates that the appendix functions as a sort of storage unit for good bacteria. Because the balance of bacterial populations inside of the gut is so critical to human health, keeping reserves on hand can give you a leg up in its restoration, should it ever get wiped out from illness, inflammation, pharmaceuticals, etc. Another theory posits that the appendix serves a more direct role in immunological functioning. Looking at the histology of the appendix, it bears tissue that looks a bit like areas called "Peyer's patches" inside the small intestine. Should the appendix prove to be functionally similar as well, it would serve as another area of immune surveillance, helping the intestines to scour the gut landscape for nefarious microbes that shouldn't be there.

Even though the appendix looks like a strange, squidgy meat pocket, it's *actually* a strange, squidgy meat pocket full of bacteria and lymphoid tissue. So, there's that.

IS THERE ANY SCIENCE TO SUGGEST THAT GUT FEELINGS ARE REAL?

Intuition is an interesting thing. The intuition of some people tells them not to take a new too-good-to-be-true job; my intuition tells me to eat chips instead of going to the gym—it's obviously a very personal and unique sensation. But is there any merit to heeding that little voice in the back of your head?

The idea of an unconscious sixth sense is incredibly difficult to quantify—how do you measure the accuracy of the way a person feels "deep down"? This has been a major barrier to studying the intuitive phenomenon through the course of modern psychology. As a result, evidence for its existence is pretty paltry.

THIS DUDE IS DEFINITELY STANDING YOU UP...

Luckily, a multi-institutional team of Italian scientists was able to reveal a possible part of the perceptive puzzle.[15] In showing groups of study participants short video clips of a hand reaching for a bottle of water, the research team asked each individual to predict if the video would end with the person drinking from the bottle or pouring it out. Surprisingly, without the aid of any distinct cues, the participants guessed correctly the action of the hand with a surprising degree of consistency—consistency that would not be able to be explained by dumb luck alone. When the study participants were asked what led them to make their choices, the majority could not give a concrete rationale.

Investigators of this research team explained that areas in the brain that are activated by physically interacting with an object may still engage even when watching someone *else* interact with the same object. This mirroring of neurological activity may subconsciously help people to anticipate actions or outcomes of events taking place around them. Small revelations, but fascinating, nonetheless.

Scientists have a long way to go in unpacking human intuition. But I'm still hopeful for the release of research findings that may help me better understand how to use my intuition to pick winning lottery numbers.

WHY DO SOME THINGS (HEART, LUNGS, ETC.) MOVE ON THEIR OWN, BUT MY ARMS AND LEGS DO NOT?

ALRIGHT HEART—HERE WE GO! 1, 2, 3: BEAT! 1, 2, 3: BEAT!

The human body operates via a complex network of peripheral nerves, all of which conduct electrical signals either to or from our central nervous system. The central nervous system, specifically the brain, is the biological version of the central processing unit in your computer—it receives various input data, processes it, and sends output signals to elicit the corresponding action. Some of these actions occur without your voluntary efforts (heartbeat, breathing, contractions of the intestine to move food through the gut), and some of these actions occur because of your voluntary efforts (reaching for a glass of water, lifting a heavy dumbbell, typing the pages of a book about science).

In order to separate involuntary from voluntary actions, the body has developed separate circuits—one is called the autonomic nervous system (involuntary actions), and one is called the somatic nervous system (voluntary actions). Each circuit has dedicated nerves and pathways, which relay signals to and from the central nervous system. The reason you don't consciously control the beating of your own heart but still control the movement of your arms is that their actions are wired via separate pathways, controlled by different combinations of neurological processing patterns in the brain.

The compartmentalization of these two systems is incredibly advantageous. It allows the body to allocate the workload of vital functions, categorized by internal and external spheres of influence. As such, the somatic system helps to perceive the *external* environment and directs body movement to engage with that environment. Opposingly, the autonomic system collects status updates on your vital, *internal* functions (like blood pressure, temperature, heart rate, etc.) and helps the brain adjust these parameters to keep you operating optimally.

Both systems are critical to survival, and giving each its own resources helps maximize the physiological efficiency along these routes of communication. Trust me, you wouldn't want to try to consciously control your vital life functions. I have a hard enough time combating procrastination as it is—I'd hate to see the dire consequences of not managing my time appropriately when it comes to consciously regulating my own heart rate.

WHY DO HUMANS CRAVE UNHEALTHY FOODS IF THEY'RE SO BAD FOR US— SHOULDN'T WE BE DRAWN TO HEALTHY FOOD?

By way of full disclosure, I have an insatiable sweet tooth. I can sit down, crush one dozen donuts, casually close the box I have desolated, and not even feel *that* bad about myself. Funny enough, this unique talent actually says more about human evolution than it does about my (occasional) poor choices.

The human body cannot differentiate "good" food from "bad" food—these are definitions that *we* have created as a society to describe what we're eating. But the human body is far more objective in its nutritional evaluations. When you eat a chocolate cake, your body sees it as carbohydrates, fats, and a tiny bit of protein. Similarly, when you eat a rice bowl with vegetables and chicken, your body also sees this as a source of carbohydrates, fats, and protein. The chocolate cake isn't a bad food, according to your body—it's a food source with an abundance of glucose.

The one thing you should know about the human body? It *loves* glucose. We have evolved to run on the glucose molecule. While we may utilize fats and proteins to aid in the manufacture or repair of certain tissue types, the energy source that fuels everything we do is glucose. If you don't eat enough glucose, your body will catabolize your tissues to make more; if you eat too much (like one dozen donuts), your body will store it for future use.

The most glucose-hungry organ in your body is the brain. While it only makes up approximately 2 percent of your overall mass, it consumes upward of 25 percent of your available glucose. So it may come as no surprise that, evolutionarily, your brain has forged addictive feedback mechanisms to the glucose molecule, driving you to keep consuming it. This means that consuming glucose physically alters the chemical cocktail in your brain—lighting up the reward center, releasing dopamine, and neurochemically encouraging you to keep consuming.

The reason we crave unhealthy food—like donuts or chocolate cake—is that our brains have programmed us to consume glucose-rich foods as a means to meet their own hefty needs. Our baked goods and delectables just happen to be fantastic sources of this molecule.

Today I realized that the human brain is manipulative and is bending me to do its bidding. It's not that I *want* to open a second box of donuts. My actions are being dictated by the devious pink spongey control freak inside my skull.

CAN YOU DIE FROM HEARTBREAK?

Heartbreak hurts like hell. When you're in the throes of sadness, it can feel like you're going to die. Coincidentally, you actually *can* die.

Sorry, dear reader. I probably should have warned you that science can sometimes be super depressing.

A clinical diagnosis associated with heartbreak is called "takotsubo cardiomyopathy." This is a condition in which the left ventricle of your heart (the chamber that pumps oxygenated blood to the rest of your body) experiences a dysfunction, leading to impaired or ineffective pumping. It presents with symptoms similar to those of a heart attack—including chest pain, weakness, nausea, shortness of breath, and even elevated biomarkers—but without any coronary artery blockages. Takotsubo cardiomyopathy is caused by an onset of sudden, extreme stress or emotional upset. For that reason, it is also called "broken-heart syndrome."

While death in these cases is incredibly rare, it *has* been reported. However, with appropriate monitoring and treatment, the condition typically resolves. Kind of like the cause of the emotional heartbreak in the first place.

(Hopefully, that last line was a little more uplifting than the first ones.)

HOW DOES CAFFEINE KEEP YOU AWAKE?

Caffeine is a stimulant, and a rather popular one at that. In 2020, the National Coffee Association (which, apparently, has been a thing since 1911—who knew?) released its results from a comprehensive survey regarding coffee-drinking habits. It was found that approximately 70 percent of American adults consumed coffee at least weekly. So caffeine consumption is pretty pervasive. Yet not many people understand the mechanism behind their cup of morning pick-me-up.

Caffeine affects the human body in a few different ways, but the most palpable contribution it makes is in keeping you awake. Although it may sound a little backward, from a biochemical standpoint, caffeine keeps you awake by preventing you from being sleepy.

The caffeine molecule fits rather nicely into the receptor site of a neurotransmitter called "adenosine." One of the tasks of adenosine is to dock into its receptors and trigger events that power the central nervous system down for rest—as neuromodulators go, it's good at slowing down brain activity. When caffeine binds to adenosine receptors, it keeps adenosine from doing its job. Thus, the sleepy signals get attenuated, and you ultimately begin to feel perkier and livelier.

Several research teams have indicated that caffeine tolerance may also be attributed to adenosine receptors.[16] When caffeine occupies adenosine receptor space, the body misses out on some of its expected sleepy-time signaling. To reestablish balance, neurons begin to manufacture more adenosine receptors to provide more chances for adenosine to beat caffeine to the binding punch. This provides more adenosine-binding opportunities, which lead to more instances of successful adenosine binding and increased feelings of being tired. As a result, many people will intake more caffeine to overcome these feelings, causing the neurons to produce *even more* adenosine receptors, and this pattern continues into perpetuity until the consumer has reached a substantial degree of caffeine tolerance.

And when I say "the consumer," I absolutely mean myself. I do this, and I am guilty of orchestrating my own detrimental need for excessive caffeine.

SPACE

"THE UNIVERSE IS UNDER NO OBLIGATION TO MAKE SENSE TO YOU."
—NEIL deGRASSE TYSON,
ASTROPHYSICIST AND BESTSELLING AUTHOR

HOW DOES GRAVITY WORK?

Hi, it's me. I've come to feed you a cosmological cop-out, so here it is: we don't really know how gravity works . . . yet.

This may not sit well with you because gravity is a force that you likely have an intrinsic understanding of. Additionally, it's absolutely critical to our existence. And I'm not kidding. Gravity is implicated in a multitude of biological processes, like keeping your bones from dissolving, helping your brain interpret the orientation of your body, and assisting the movement of broken-down food as it travels through your gut-pipes.

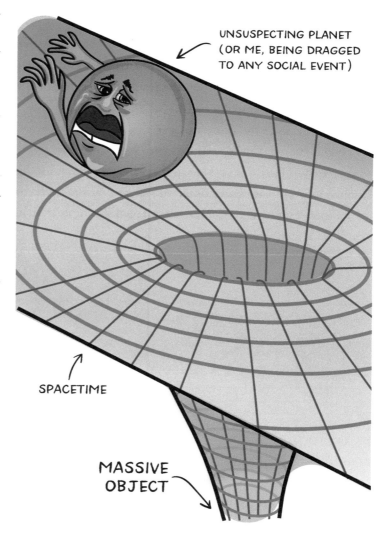

(Sidenote about that last bit: gravity is vastly important for pooping, from both a physiological and performative standpoint. Orbiting astronauts have a harder time with, well, making bowel movements. Without gravity to pull everything downward, gut motility relies on muscular contraction alone. As a result, our brave men and women can become constipated. Early on in the space program, there was also a sharp realization that making boom-boom without gravity complicated the . . . disengaging process as well. Again, without anything to pull downward on the stuff that comes out, it kinda hung out near the exit, if you will. So the world's brightest engineers came together and invented a high-powered poop vacuum. Which was modified many times following dire space failures. And it is still proudly used today.)

So, we know about gravity—we can see what it does, we can feel it, we can measure its force—but we still haven't confirmed, for sure, how it works. The most accepted macroscopic theory for how gravity *might* work comes from our good buddy Albert Einstein via his general theory of relativity. While it is chock-full o' math, the theory basically states that massive objects warp the fabric of spacetime. Not unlike tossing a bowling ball onto an outstretched sheet (the bowling ball = a massive object; the sheet = spacetime), the resultant warping of spacetime creates a kind of funneling effect that may drag nearby objects in.

Bottom line: the more massive the object, the harder it warps spacetime and the larger, more powerful funnel it makes! This is why something like our sun—which is about 333,000 times more massive than Earth—can keep eight planets (and an asteroid belt, and the Kuiper Belt, and the Oort cloud) locked in at extraordinary distances.

Pluto stays loyal to the sun, despite a 3.7-billion-mile long-distance relationship . . . and you still can't get a text back.

WHAT DOES SPACE SMELL LIKE?

I probably won't blow your mind when I say this: space is a big dark vacuum.

I probably *will* blow your mind when I tell you this, though: space is far from empty.

The inky blackness of space contains a suspension of powdery dusts, bursts of exotic particles, and an array of radiation from the electromagnetic spectrum. These bits of mass and zippy charges do have a detectable scent. So what does a whiff of space smell like? The following are actual astronaut descriptions:

- "WELDING-TORCH FUMES"
- "SWEET METALLIC"
- "GUNPOWDER"
- "SEARED STEAK"
- "HOT METAL"

(In reading this back to myself, I'm struck by how much it sounds like some bizarre, unnerving, postapocalyptic grocery list.)

There seems to be a commonality among the scent descriptions: they all sound like the throbbing iron components in some large machine shop. So I think this begs the obvious question: why? Well, I hope it's painfully obvious to my readers that astronauts do not smell space by popping off their helmets—midspacewalk—and inhaling deeply through their nostrils.

(That *is* understood, right? Space is a vacuum, and there's nothing to breathe, and the saliva would boil inside of your mouth as it turns into a gas . . . Okay, cool. Just checking.)

Astronauts detect the smell of space after they reenter their habitable spacecraft. As they climb back inside their cosmic vehicles, the small bits of space stuff cling to their suits. When they remove their helmets, they can smell the celestial particulates they dragged in with them. It's thought that the reason it smells metallic is due to the chemical reactions that take place inside the spacecraft between the stuff from space and the artificial atmosphere inside. These are likely oxidation reactions (the same chemical dance that causes the rusting of metals, among many other things).

Speaking of chemistry and olfactory senses and space! Scientists recently discovered that an area at the center of our galaxy is packed with clouds of ethyl formate. Coincidentally, this compound is also what gives raspberries their distinctive, delectable taste. Which confirms the theory that the Milky Way is actually just a very large berry-filled tart.

HOW DO WE KNOW THE UNIVERSE IS EXPANDING?

The universe is impossibly large. So large, in fact, that we cannot see all of it. This is why you may have heard astronomers and astrophysicists coyly drop the term *observable universe* as another way to humbly say, "Yyyyyyyeahhh, it's pretty massive out there . . . could even be infinite . . . and what I'm about to describe is limited by how far our instruments can reach." Currently, the area of the universe that we *can* see is about ninety-three billion light-years across. In other words, if you were to travel at the speed of light, it would take you ninety-three billion years to travel from one end to the other.

So, you know, pack a snack.

Considering these incomprehensibly large distances, we obviously can't keep track of the changing size of the universe with a tape measure. But we can observe light from our neighboring galaxies to find out if they're hurdling toward us or rapidly moving away. This may not sound like a reliable system if you're unfamiliar with electromagnetic wavelengths. However, it's pretty easy (relatively speaking, from a physics standpoint) because of a phenomenon known as the Doppler shift.

Bear with me here.

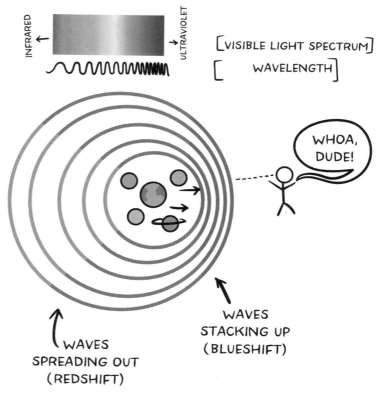

Doppler shift is the principle that describes the way electromagnetic wavelengths change, based on the relative motion of their source. If a source of radiation that you observe is moving toward you, the wavelengths get shorter: this is called "blueshifting" (because blue-violet light is associated with short wavelengths). If a source of radiation is moving away from you, the wavelengths get longer: this is called "redshifting" (because red light is associated with long wavelengths).

Now, you likely haven't *seen* Doppler shift to confirm these points—your friend may have run away from you, but he certainly didn't appear to turn red (unless he was super angry). But! I guarantee that you've *heard* Doppler shift, which also applies to sound waves. If you've ever stood on a street corner as an ambulance drove by, then you're already an expert:

You hear an ear-shredding siren, blaring on the next block over. The ambulance heaves toward you, closer and closer. As it nears your location, you notice that the pitch of the siren gets higher.

The ambulance passes, and a warm air mass washes over you.

The ambulance is now hurtling away from you, and you notice that the pitch gets lower as it drives off.

The whole thing kinda sounds like, "weyy-oh, weeeEEEEEE OOOoohh, wahhh-uhhhhh"—a steadily higher pitch until it passes, then the pitch dives down again. High pitch = high-frequency waves, stacking up as the source approaches; low pitch = low-frequency waves, spreading out as the source gets further away.

When we observe our cosmic neighbors, we see a lot of redshifting. This indicates that almost everything around us seems to be leaving the area. Furthermore, there also seems to be a relationship between the distance of the galaxies observed and the amount of redshift they exhibit. Based on these observations, we conclude that the fabric of the universe, in which everything is suspended, is expanding at an accelerating rate. Like marker dots applied to the surface of a flaccid balloon: blowing up the balloon results in each dot moving away from all the others. Based on the fact that the galaxies and large cosmic bodies around us demonstrate a similar pattern of movement, we conclude that our Great Big Universe Balloon is expanding around us as well.

. . . the real question is: who or *what* is blowing it up?

WHAT ARE SATURN'S RINGS MADE FROM?

I'd argue that, apart from Earth, Saturn is probably the most recognizable planet in our solar system. But despite its level-A rock-star status, the ring composition is pretty elusive to most people—this curious quandary has been asked of me several times.

The direct answer: the rings are made of a multitude of things! The detailed (and arguably more interesting) answer: the rings are made of ices, dusts, small space rocks, and bigger rocks that crash into other rocks to make more small space rocks.

Saturn's rings are dynamic. The stuff that comprises them is continuously smashing into itself, lumping together like balls of clay (a process called "accretion"). When those large bodies (some of which have been observed to be the size of houses) careen into one another with sufficient speed, they shatter and return to their teeny origins. How teeny? Certain regions of Saturn's rings contain particles the size of grains of sand.

Now, the next question is obvious: where did these mesmerizing rings come from? Well, what you should know up front is that Saturn is enormous. It's one of the gas giants in our solar system, and it checks in with a hefty mass nearly one hundred times greater than that of our piddly Earth. And if you've read the entry on gravity, then you already know that Saturn probably exerts a rather large deformation on the spacetime around it.

At some point during the formation of the solar system, Saturn likely snagged rogue asteroids, minimoons, and other chunks of space debris that were unlucky enough to wander too close. Under crushing gravitational forces, these celestial bodies were likely whipped around the planet, then pulverized and strewn about, only to have their remnants locked within orbit. Over time, the resultant gravity-forged carnage spun itself into delicate rings of varying density.

In totality, the rings are about 170,000 miles across and only thirty feet high. Kind of like a cosmic thin-crust pizza. If that pizza were made from space rock and frozen ammonia—which, as I understand it, Pizza Hut recently discontinued.

WHAT IS THE SUN'S FUEL SOURCE?

On Earth, the vast majority of our combustible fuel sources are derived from complex molecules—fossil fuels, dried timber, farts (oh, come on—don't act like you haven't lit a fart ablaze, or have at least seen a friend do it). So you might be surprised to learn, dear reader, that the enormity of our sun's power relies on the simplest element in our universe's arsenal: the hydrogen atom.

Before the formation of planets, and even before our sun ignited, our infant solar system was just a large cloud of dusts and molecular gases. One particular area was especially dense and full of hydrogen. Slowly, but very surely, these hydrogen atoms came together, attracted by each other's tiny gravitational fields. As this coalescing, swirling area became denser and denser, it attracted more of the surrounding hydrogen toward it, increasing its mass and subsequently increasing the pressure at its center. This process continued for tens of millions of years, creating a super dense ball of gas on the verge of bursting into a brilliant space flame.

Eventually, these heaving, inward pressures reached a tipping point: the crushing pull became so powerful that it actually mashed the nuclei of atoms together, merging them into one. This is called "nuclear fusion." When two hydrogen atoms (each of which has one proton) combine, they form a new helium atom (which has two protons). But the energy required to merge atomic nuclei is staggering, and thus the resultant energy release is correspondingly powerful. Every time hydrogen atoms fuse, a burst of heat energy, visible light, and other electromagnetic radiation is burped up. This energy release is then radiated outward. It's also the same heat, light, and radiation that we experience from our sun while napping on the beach on Earth.

It's worth noting that nuclear fusion is no small feat. The kind of pressure and temperature you need to accomplish nuclear fusion, from a physics standpoint, is ridiculous. By way of example, the temperature at the center of our sun, which is a gigantic fusion reactor, is approximately 27,000,000°F; the pressure at its center is 3,800,000,000 psi. As the sun's core fuses hydrogen, the resultant energy release helps to keep conditions hot enough to continue the great big chain reaction of fusion. It is this chain reaction—fusion, energy release, maintenance of extreme temperatures, more fusion, more energy release—that has been ongoing for billions of years.

Eventually, the sun—like any star—will convert all its internal hydrogen stores into helium. At this point, without reliable fuel for fusion, it will begin its death process. Which is incredibly dramatic and involves the sun expanding rapidly, eating some of the interior planets, then getting super tired and shrinking down to a pathetic, burned-out matchhead. Luckily, it is only about halfway through its life, and probably experiencing some kind of midlife crisis. Like going shopping for a Porsche it can't afford. Or dating a much younger star.

IF EARTH'S GRAVITY IS PULLING ON THE MOON, WHY DOESN'T IT CRASH INTO US?

Simple question with a math-heavy answer. Which is actually quite common when you dive into physics. Which I learned the hard way . . . when I was at Harvard . . . and I waited until the last minute to complete a physics homework assignment . . . and it took my partner and me three hours to solve a single problem . . . and I'm *absolutely* sure that we still answered that problem *absolutely* incorrectly.

Anyway, my past academic physics trauma aside, I think I can give you a solid rundown of this concept. Hell, I'll even spare you the tedious calculations.

It goes something like this: the moon *is* falling toward Earth, but it just keeps missing.

"Leah . . . what???"

I know. But the crux of this principle applies to all orbiting bodies (including the International Space Station and—as of the writing of this book—6,542 satellites). Now, due to the attractive force from gravity alone, with no other motion, the moon and Earth would be drawn directly toward one another (which would end in a disastrous planetary hug). Luckily, the force of gravity is *not* the only component which dictates moon motion in this curious equation: the moon also exhibits a superfast orbital velocity (we say that this motion is *tangent* to it's circular, orbital path).

Think of it like this: If you were to stand in the middle of a field and throw a ball out in front of you, the ball would cover a certain amount of distance before gravity pulled it down—the overall ball path, from launch to landing, would form a large arc. If I were to tell you to throw the ball harder (like my aggressive eighth-grade gym teacher), the ball would cover even more distance before being pulled down by gravity, forming an even larger arc.

Now, let's say that I teleported you, with your ball, to the top of Mount Everest. I tell you to throw the ball again as hard as you can, over the side. Once your ball hits the ground, the actual ground-level distance it covers would be significantly longer than ever before and its arc-path would be gigantic. If you could keep getting higher above Earth's surface, and keep throwing the ball with more and more velocity, eventually the distance of the ball's trajectory would be so far that its arc would match the curvature of the Earth. This means that the path of the ball you threw would never hit the ground—it would completely encircle the planet! So once you pitch that superfast, superlong throw that breaks

all kinds of world records . . . the ball would just keep circling over and over and over again. This is why our moon never crashes into us—while it is being tugged toward us by gravity, it's also traveling so fast, laterally, that it keeps missing.

You've got to be pretty high above the ground, with a sufficiently fast speed, to accomplish orbit. For stable-orbit rocket launches, during the final stage of flight, this shakes out to about 7 km/s (20 times the speed of sound or so), after having accomplished something called a "pitchover maneuver," which slightly changes the upward trajectory of the initial part of the launch into more of a lateralized arc above Earth's surface. Similar to the ball throw.

Sidenote: If you *do* end up tossing a ball into orbit, there's probably a hefty MLB contract waiting. And I'm pretty sure I'm entitled to some kind of finder's fee for referring you.

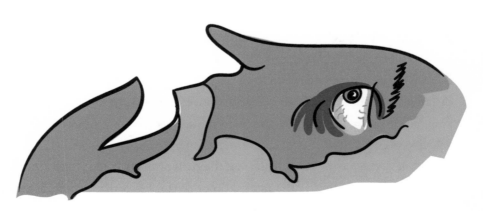

IS THERE ACTUALLY A HOLE IN THE OZONE LAYER, AND WHAT IS IT FROM?

There *is* a hole in the ozone layer. The explanation for this involves an intricate chemical waltz. But first—the basics!

What is the ozone layer? The ozone layer is a section of our atmosphere, served with a generous helping of the molecule O_3 (this is what we call "ozone"—three oxygen atoms attached in a small daisy chain). The abundance of this little molecule directly impacts our health. You see, ozone serves to block the UV radiation from the sun—one of the nasty parts of the electromagnetic spectrum that causes skin cancer. So you can think of the ozone layer as an atmospheric application of sunscreen: it keeps us from becoming seared little meat pieces. As such, keeping it intact is pretty important.

When scientists say, "a hole in the ozone layer," they are referring to an actual, measurable chunk of this protective barrier that is exceptionally thin or missing. And it's not small: as of 2021, this gaping maw was reported to be larger than the continent of Antarctica. But there isn't a gigantic ladle steadily scooping O_3 out of the atmosphere. The reason for the defect in our ozone layer comes down to dirty chemistry, fueled by man-made industrial by-products.

HOW IT'S SUPPOSED TO WORK: As the sun's UV radiation rains down onto the stratosphere, oxygen molecules split into two solitary oxygen atoms (because they have a rogue electron, we call them "oxygen radicals"). But these lone atoms are single and ready to mingle. In a healthy ozone layer, the oxygen radicals then rapidly recombine with molecular oxygen to form ozone. At this point,

> SON OF A...HOW LONG HAS MY FLY BEEN OPEN?!

ozone absorbs more UV radiation and breaks apart, yielding more oxygen radicals, which recombine with molecular oxygen to make more ozone, ad nauseam.

It's a big self-perpetuating cycle—called the "Chapman cycle"—that kinda looks like this:

$$O_2 + \lightning = O\cdot + O\cdot \text{ (Radiation breaks molecular oxygen apart to form oxygen radicals)}$$

$$O\cdot + O_2 = O_3 \text{ (Molecular oxygen and oxygen radicals form ozone)}$$

$$O_3 + \lightning = O2 + O\cdot \text{ (Radiation breaks ozone apart to yield more molecular oxygen and oxygen radicals)}$$

$$O_3 + O\cdot = O_2 + O_2 \text{ (Ozone combines with oxygen radicals to form two molecular oxygen atoms)}$$

You don't really need to understand all of these chemical equations to get the concept—just know two important points: (1) ozone and oxygen recycle into each other in a big ol' harmonious equilibrium, and (2) the chemical process of splitting and reforming ozone and molecular oxygen is just a complicated way of displacing and absorbing UV radiation before it reaches Earth's surface.

HOW IT'S NOT SUPPOSED TO WORK: Now, remember what I said about oxygen radicals being ready to mingle? Well, oxygen doesn't discriminate when it comes to dance partners. Which is okay . . . until human beings release something called "chlorofluorocarbons" (CFCs) into the atmosphere. These molecules are used in industrial refrigerants and solvents and as propellants in aerosol cans. They contain the atoms chlorine, fluorine, and carbon (hence the name). Now, when CFCs are released by industrial processes, they float upward through the atmosphere. When they reach the stratosphere, UV radiation breaks them apart, yielding chlorine atoms. These chlorine atoms bonk into ozone molecules and rip oxygen atoms from them, creating regular molecular oxygen. At some point, the chlorine will let go of its oxygen and then zip around, breaking apart more and more ozone.

These CFCs directly disrupt the self-perpetuating cycle we (laboriously) talked about above. With enough of these chemicals, a measurable (and super large) hole is made in our ozone armor.

The ozone hole is seasonal and located over the South Pole. The reason for this is due to a myriad of climate conditions, atmospheric chemical composition, and global weather patterns. But we can be sure of one thing: thankfully, Santa Claus is spared exposure to intense bursts of UVB light. He only has to worry about his house flooding as the northern polar ice cap melts . . . also because of humans. Go us.

IS IT TRUE THAT ALIENS LIVING SIXTY-FIVE MILLION LIGHT-YEARS AWAY WOULD STILL SEE DINOSAURS IF THEY LOOKED AT EARTH?

If you've never heard of this phenomenon, then you're probably reading the question again and again, thinking, "What the fffuuu . . . ?"

The cool thing? The answer is yes. And I'm going to tell you why.

(A slight nuance for this entry: sixty-five million years ago is approximately when the dinosaurs went extinct. So, for the sake of argument here, let's just say that this extinction event had not yet happened, from the relative perspective of these stargazing aliens).

The power of sight is directed by the almighty photon—this is the quantum particle of the electromagnetic spectrum. If the visual-sensory system of these aliens is similar to ours, then they have some kind of eye structure with a retina at the back. Now, the retina is a layer of tissue containing specific cells (called "photoreceptors") that become activated when they come into contact with photons. This activation culminates in a signal that is sent through the optic nerve to the brain for processing and visual interpretation. Voilà! Sight.

Okay, prepare your mind, for it is about to be blown.

A light-year is a unit of measurement. It is the distance that you would cover if you traveled at the speed of light for one year. So if our voyeuristic alien neighbors lived sixty-five million light-years away and we wanted to visit them, it would take us sixty-five million years to get there if we traveled at the speed of light.

Stay with me.

As I explained above, our power of vision is guided by how photons activate our brains. Photons enter our eyes; we interpret them and form images in our heads. So! Photons that were emitted from our Earth (bouncing off dinosaur scales, perhaps) would race away from Earth at the speed of light. After sixty-five million years, they would just now be arriving at our alien friends' planet, sixty-five million light-years away from us! And the aliens would, indeed, see the light that has traveled from our ancient past (because the photons being emitted from our present day are still en route).

Now, it's unlikely that this alien civilization will have the ability to see our Earth's surface with enough resolution to appreciate individual dinosaurs. But, with enough advancement, they might certainly be able to see what our solar system looked like sixty-five million years ago.

This concept resonates in the cosmos, today, from our vantage point. Some of the stars in our treasured Hubble images have since exploded and no longer exist in the blazing form that we currently observe. Simply because the photons from their cataclysmic ends have yet to reach us.

So, yes, aliens that live super-duper far away can still see stuff that's in our super-duper past. Which is probably a good thing: it would theoretically be way cooler to peek through a futuristic telescope and see gigantic lizards, rather than to watch some squishy middle-aged dude mow his lawn.

WHAT THE HELL IS DARK MATTER?

I'm going to start this off by telling you that the science of the cosmos is both exotic and elusive. As we continue to evolve as basic organisms—on our humble iron rock in the middle of space—we develop increasingly precise equations and more powerful instrumentation to better understand our dynamic universe. But overall, we still don't know much. And many of the explanations to questions like this one go something like this: "Well, we don't know exactly *what* it is, but we know how it might work . . ."

Dark matter is a heavy-handed suggestion—it's something we can't see, but we observe what it does to everything else we *can* see. It is, primarily, based on the otherwise-unexplained movement of large bodies in space and the observation that, without its existence, many of the important clumps of mass, which form the webwork of our known universe, would cease to exist.

The concept is pretty simple in theory. Gravity is the attractive force that holds mass together. This attraction is what maintains the planetary orbits around the sun, what keeps your feet on the ground, and what prevents the molecules in your body from blowing apart. And while gravity between massive objects (think stars, planets, etc.) can act at astounding distances, the universe is so incomprehensibly big—and the gaps between celestial bodies so enormous—that there is little chance many solar systems, or galaxies, or superclusters of galaxies, could generate enough gravitational force to be held together *only* with the force from the matter we can measure.

Basically, as we look out into the inky black of space, we can't find enough stuff to generate an adequate amount of gravity for everything to stay mashed together or behave in the way that it does. Additionally, astronomers have found areas of unexplained gravitational lensing—warpage of space-time that is so severe that we witness light bending around some unknown and currently undetectable, lurking mass. Ready for me to rattle your noodle even further? Based on the careful, collective calculations of our best astrophysicists, the concentration of dark matter must be about five to six times greater than atomic matter for our physical observations of space to make sense. In other words, the mysterious substance that we can't directly see or readily describe? Yeah. It occupies way, way, WAY more room than the matter we're familiar with.

UNIVERSE SCORECARD:

Regular matter (the stuff that makes up you, me, your neighbor Doug, planets, and stars): 5 percent of the total universe.

Dark matter (the ghostly substance that exists all around us, that we've only recently been able to conceptualize): 27 percent of the total universe.

IF NO LIGHT CAN ESCAPE BLACK HOLES, HOW DO WE KNOW THAT THEY EXIST?

As far as I'm concerned, there are only two universal certainties: (1) never start a land war in Asia, and (2) the gravitational pull of a black hole is so extreme that not even light can escape it.

If you've already read the entry on dinosaur-gazing, voyeuristic aliens, then you have a solid understanding of how photons activate the retina to provide us with the sensory capability of vision. And as a student of science, you then intimately understand that if no photons are escaping from black holes, then there are certainly no photons entering your eyes, and therefore you would not be able to see the black hole itself.

You make me very proud, dear reader.

So, how do you go about proving the existence of something without direct visual confirmation? In science, we use the same foundation for every major theory: pitch a harebrained hypothesis, get a lot of crazy-brilliant peers to test your idea for flaws, then draw appropriate conclusions. With respect to black holes, the conceptualization of their existence came down to brilliantly complicated math (I'm talking pages of calculus, folks—I have literally had nightmares about derivates like this) and eventual confirmation of their presence by observing the consequences of what their extreme gravity fields can do to matter around them.

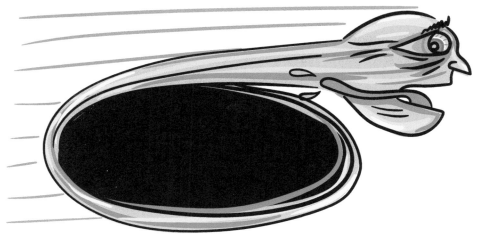

The original framework that predicted the existence of black holes came from Albert Einstein in 1915 (big surprise, right?). This prediction was embedded in his general theory of relativity, which, to jumpstart your memory, explains gravity as a warpage of spacetime around massive objects. For many decades following his publications, scientists from around the world steadily chipped away at the idea that hypermassive objects could theoretically bend the fabric of spacetime to such an extreme degree that nothing could escape the steep descent into the void they create. The edge of this gaping space-maw is called the "event horizon"—a threshold past which nothing, not even light, can return.

It wasn't until the 1960s that astrophysicists began to collect the first real-world data to begin to substantiate Einstein's (and later his colleagues') postulates. Years of data collection indirectly detected black holes via the observation of powerful X-ray bursts, watching stars whip around invisible gargantuan partners, and observing other stars being pulled out like glowing, celestial taffy. These phenomena could only be driven by incomprehensibly strong gravitational forces, derived from some unseen hypermass: *ding-ding* a black hole.

Finally, in 2019, a brilliant team of researchers synthesized data from the Event Horizon Telescope to directly photograph a legendary void.[17] This historic image shows the shadow of a supermassive black hole, surrounded by glowing plasma, right smack in the middle of a galaxy called "Messier 87." A couple of years later, the team was able to further refine their image to reveal magnetic-field lines carved through the circumference soup of plasma, as well.

All of this being said, we know black holes exist because the theoretical physics predicts it, the indirect observations uphold the physical principles described, and—oh yeah—we've now been able to capture an image of the monster's shadow.

Fun fact: black-hole gravity is so strong that it can pull matter out into long threads. The technical term for this is "spaghettification." And I promise I'm not making that up. Scientists actually present their research at important conferences, in front of illustrious experts, using this term to basically say that they observed whole stars being turned into space pasta.

WHY ISN'T PLUTO A PLANET ANYMORE?

Based on universal principles that describe the laws of nature, scientists can predict innumerable biological, chemical, and physical outcomes, given a few bits of background information

That being said, I can assure you—with absolute certainty—that they could have *never* predicted the booming public outcry after Pluto was officially demoted to "dwarf planet." In fact, planetary scientists drowned in letters from outraged people and were lambasted on internet forums around the world. One scientist from UCLA even claimed to have received an email-based death threat!

You see, in 2006, planetary scientists and astronomers agreed that Pluto did not meet the criteria necessary to maintain its status as a bona fide, varsity-level planet. These criteria include:

- ORBITING THE SUN (CHECK),
- HAVING ENOUGH MASS THAT IT CAN MAINTAIN A SPHERICAL SHAPE (CHECK), AND
- HAVING ENOUGH MASS THAT IT HAS CLEARED OUT THE AREA AROUND IT OF SMALL STRAY NEIGHBORS (FAIL).

It turns out, Pluto simply does not have the gravitational gumption to rein in the debris in its swath of space. By comparison, each of the other eight planets in our solar system has already done this—vacuumed out their areas via their own gravity, leaving only their companion moons behind. This is an important distinction: if you're big enough, and established enough, you *should* be the top dog in your territory and not just another piece of wandering space rock.

Pluto also just so happens to orbit in and out of an enormous band of frozen space rubble called the "Kuiper Belt" (pronounced *kai*-per). This belt includes several Pluto-sized miniworlds that have already been classified, with thousands upon thousands more yet undiscovered. Between its failed test of planetary criteria, tracking Pluto's wonky trajectory, and noting that it *kinda* looks like the other rocks out there, scientists decided that Pluto simply wasn't special enough to be a real planet anymore. It seems more likely that it was misclassified from the jump. And thus began Pluto's highly contested fall from greatness.

Pluto: ninth from the sun, but (apparently) first in the hearts of angry letter writers.

WHAT ARE ORGANIC MOLECULES, AND WHY ARE WE LOOKING FOR THEM ON MARS?

When chemists use the word *organic*, they are not referring to a specialty process of synthetic herbicide-free agricultural cultivation.

In chemical terms, an organic molecule consists of carbon atoms connected together and linked to other elements like nitrogen, and/or oxygen, and/or hydrogen. We label these molecules as organic because they are the materials that comprise the building blocks of life as we know it.

At the time of the drafting of this book, the *Perseverance* rover—part of the Mars 2020 mission—is galivanting about in the rusty Martian soil. One of its many objectives is to sample the alien dirt for these organic molecules. If present, they would be a strong indication that the currently barren (at least, from outward appearance) planet potentially harbored life at some point in its distant past.

This previous sentence probably isn't punchy enough to fit the *absolutely f***ing incredible* discovery that this would be. It would, without a doubt, be the most important finding since early humans grunted their way through the accidental discovery of fire. The discovery of previous life on Mars would be a small step in peeling back the curtain that conceals the answers for where we potentially came from and, even more provocative, whether we're alone out here.

Now, the reason we're sampling for organic molecules as an indication of life is simply that we don't know what else to look for. Our myopic concept of what constitutes "living" is colored by our singularly unique experience on our planet—living things here contain these kinds of molecules. Could there be alien life with a drastically different chemical combination? Sure. And if I'm giving you my personal opinion, I'd say there is a 100 percent chance that there probably is. But we can't find what we're looking for if we don't know what we're looking for—get it? So, in science, we start with the fundamentals—what is it that we *know* about life here that we could potentially apply to our all-important search? And that's where scanning for these compounds comes in.

In my career, I settled into medical sciences. I consider myself unbelievably fortunate: I love biology and pathophysiology, and I get to apply my background in medical research and biotechnology to help to solve some of our terrestrial, human problems. But I have always said that if I had to do it again, I'd dive into a space focus; I'd want to contribute to the quest to confirm the existence of life elsewhere. However, I would have strongly suggested not allowing me to do any initial introductions with any intelligent life-form. Because I'm terrible with names—I would undoubtedly forget Xendorph-6's name *immediately* after meeting it.

WHY CAN'T ANYTHING TRAVEL FASTER THAN THE SPEED OF LIGHT?

In a universe teeming with boundless physical possibilities, it seems strange that the cosmos has some kind of no-questions-asked speed limit. And that speed limit is about 300,000 kilometers/second, or just under 671,000,000 miles/hour (for my American readers). Now, there are two answers to this question, depending upon whether the traveling object under consideration has mass or not. Each scenario bears slightly different rationale. But for the sake of simplicity (ha! "simplicity"—there's really nothing simple about this), let's assume that we are talking about a massless photon.

The theory behind this strictly enforced velocity includes a phenomenon known as "time dilation." Now, for this exercise, I'm going to need you to think outside of the Earth-bound box. Ready? Deep breaths . . .

Rather than thinking of time and space as different entities, try to consider them as intimately connected. Space is tangible—we can experience it in three distinct dimensions (up-down, side-to-side, and forward-backward). But I want you to tack on a fourth dimension that is *just* as tangible: time. Toss your wristwatches out the window—for the sake of this analogy, time is no longer numbers on a dial; it is now a coordinate with directionality.

Okay, here we go: let's say you're driving a car. Your destination is somewhere northeast of your current location. To get there, you can start by heading due north, due east, or some combination of the two, right? If you head due north, there is no part of your car's movement that is headed east—it's going entirely north. Similarly, if you head due east, there is no part of your car that is headed north—it's going entirely east. If you head northeast, you're moving a little north and a little east at the same time.

Easy.

Now, to construct an exceptionally basic analogy for time dilation, we replace east with space, and north with time. Nothing changes in our analogy. Just the names.

Using our newly-labeled spacetime axis, as you drive around in your car, you're traveling a certain distance through space over the course of a period of time. So you're headed through some of space and some of time simultaneously (or northeast, if applied to the original example). Using the exact same principles in the example above, if you *only* head through time (north), you are moving through none of space (east), and you'd be at a dead stop. Similarly, if you *only* head through space (east), you are moving through none of time (north), then you arrive at your destination instantly. Time dilation dictates that the faster and faster you travel, the slower that time (relative to you) becomes; as you approach the speed of light, your relative time grinds to a halt.

For photons in the vacuum of space, they max out the speedometer of the universe. So they travel only through distances in space without a time component. Covering an immense distance in zero time means that their velocity is as lightning fast as it gets without being called "infinity fast." In our universe, infinity fast = the speed of light.

The reason that photons can't travel faster than that? Because light speed is the absolute fastest that physics will permit them to go. It is instantaneous travel between origin and destination. It *is* due east. You can't get to your destination faster than instantaneously. Wanting to go faster than the speed of light is like being at a complete stop and somehow wanting to move slower—can't go slower than stopped; can't go faster than light speed.

Why does this speed shake out to about 300,000 km/s? Scientists aren't sure yet. But it's calculable, and that's just what it is.

You might have to read this a couple of times. But I promise it will click. When it does, you will rejoice. And then I will tell you that distant galaxies at the edge of the universe actually appear to violate everything you just worked so hard to conceptualize. Because light speed seems to only apply to things moving *through* the universe, but does not really apply to the layered expansion of the fabric of the universe itself. Therefore, due to the accelerating expansion of spacetime, these galaxies are probably careening away faster than our Grand Speed Limit.

Ugh. Theoretical physicists, am I right?

CAN MOONS HAVE RINGS OR THEIR OWN LITTLE MOONS?

I'm struck by how adorable the thought of a minimoon is.

Moons actually *can* have their own little moons! While we have yet to directly observe this phenomenon, it is certainly not outside of the realm of possibility. And we also just haven't had the luxury of observing exoplanets (planets outside of our own solar system) close-up, due to our current technological limitations. So the fact that we haven't seen a minimoon yet doesn't mean much.

The theory of a moon having its own teensy moon is supported by the analogy of the sun-earth-moon orbital combo. Earth is a satellite of the sun; the moon is then a subsatellite that orbits Earth. If

you were to scale this model differently to an earth-moon-minimoon orbital combo, then the moon is a satellite of Earth, and the minimoon would be the subsatellite. Not beyond the realm of possibility.

While there are a few variables that determine how satellites are snagged into orbit, the physical principle that allows for possible minimoons is called a "Hill sphere." It's a gravitational threshold that surrounds bodies in space and helps determine the likelihood of a smaller body capturing a satellite, especially when in the presence of other larger bodies. If a satellite (a moon, or a minimoon, or *mini*-minimoon) has an orbital pattern within the Hill sphere of a larger object, it will orbit that object. If a satellite has an orbital pattern outside of the Hill sphere of an object, it will likely orbit the next closest large object in the system. The Hill sphere is basically just a distance-dictated zone to determine which celestial body's lasso has the best chance at wrangling a meandering satellite.

I'd like to take this opportunity to be the first scientist in the history of the world to end an explanation of the principles of the Hill sphere with this: Yeehaw.

IF THERE WAS A BIG BANG, DO WE KNOW THE LOCATION IN THE UNIVERSE WHERE IT STARTED?

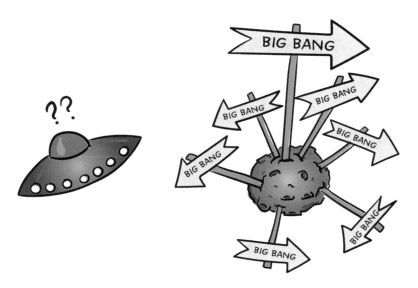

Buckle up, this is going to get esoteric.

The big bang didn't really happen in a singular, discrete location—from our perspective, it happened everywhere, simultaneously.

[dodges thrown shoe]

I'd like to take a moment to call attention to the qualifying words within that statement: *from our perspective*. Now, you're going to have to get really comfortable with something that may make you feel quite uncomfortable. The fact of the matter is that the universe operates in colors, dimensions, vibrations, and quantum states that we simply don't know about or can't conceptualize based on our five meager senses. Our hard-coded physical limitations obligate us to a very diluted perception of what may otherwise be a very complex and full-bodied universe.

When we evolved on our small space rock, we did so with physical abilities that allowed us to exist on our terrestrial world, with just enough sensory input to make our existence possible. We developed photoreceptors to see our food and to detect friend from foe (for the record: we don't even have the most complex eyes in the animal kingdom—even some variations of shrimp can detect a multitude of colors beyond our basic ROYGBIV). We can hear, and taste, and touch. But our life experience is contained within the three spatial dimensions we can perceive: up-down, side-to-side, and forward-backward.

Unfortunately, that probably isn't the extent of reality.

So back to the big bang. From where we're sitting, other celestial bodies (other galaxies and the like) are rapidly accelerating away from us. And the ultimate kicker? If you were to shuttle yourself to a distant galaxy, all its space neighbors would be hurtling away from it as well. When everything is expanding (as we have confirmed the universe is doing), it is impossible to pinpoint where the epicenter of a big bang would have been. Why? Because it likely exists in a dimension outside of our basic three.

The best analogy I can give you is that we exist on the surface of a balloon covered in dots—each dot represents a galaxy, and the flaccid latex balloon is the fabric of spacetime as we perceive it. As air is blown into the balloon, it expands. From the perspective of every dot, all the other dots around it are expanding outward and away. Now, if you were the tiniest, flattest, two-dimensional creature that lived on one of those dots, you could only perceive that the other dots were accelerating away from you. However, you wouldn't have the ability to detect the underlying expansion mechanism: the increased air pressure from inside of the balloon itself—that's a third dimension, and as a two-dimensional entity you simply don't have the sensory equipment to see it.

Back to real-world applications: As three-dimensional creatures, we can detect the movement of things suspended in the expanding spacetime around us, but we cannot trace the big bang back to a coordinate, or a quadrant, or a geographic epicenter. Why? Because it probably exists in a dimension outside of our perceptive power.

Now, I think the confusion comes in with the misnomer: "the big bang." We're familiar with Earth-bound explosions here, and they can always be traced back to a singular locale. But the birth of the universe was not an explosion per se: it was a nonexistence that sprang into existence, everywhere. At least, as far as we can see it. So it's more like the Big Everywhere-Expansion-from-Nothingness-All-at-Once.

If you hate this answer, that's okay. Because it's located most of the way through the book. So you've already been obligated to read this far. And that's not a coincidence—that's just damn good planning on my part.

IS THE MILKY WAY SIMILAR TO OTHER GALAXIES?

NOT A SPECIAL SNOWFLAKE

I know what you want to hear. This is one of those rhetorical questions in which the asker is only seeking one particular response—kind of like the space version of "Does this outfit make me look fat?"

Pleasant lie: no, the outfit does not make you look fat. Unpleasant truth: yes, the Milky Way is similar to other galaxies.

As sentient creatures aware of the deep expanse of space around us, we possess a uniquely human hubris. After all, we are the only planet that harbors life in this part of the universe (at least, as far as we currently know). So there's a special searing pang of insecurity when we hear that our galaxy is *not* especially unique.

In general, there are three major categories of galaxies based on their observable shape: spiral, elliptical, and irregular. Now, these categories have subdelineations (like barred spiral or interacting), but for our purposes, we can stick with the Big Three.

The Milky Way is a spiral galaxy. Remember what I said about humans wanting to feel special? Well, 60 to 70 percent of galaxies within the observable universe are also classified as spiral shaped. Many of these galaxies, even in our immediate vicinity, bear a significantly higher amount of brawn than we do. In fact, our closest neighbor, Andromeda, is approximately *double* the width of the Milky Way; others, like IC 1101, may be up to sixty times larger! Therefore, not only are we similar to most other galaxies in terms of shape, but we're also not super impressive in size.

So the next time you feel grossly unappreciated at work? Just remember you're also residing on a small rock orbiting around an average star in a nondescript outer arm of a midsize galaxy, which could be mistaken for billions of other galaxies.

#inspiration

SPACE

HOW MANY GALAXIES ARE IN THE UNIVERSE?

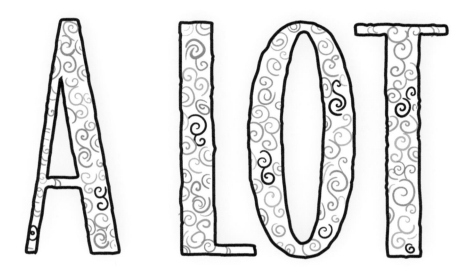

The universe is big. *Really* big. And its diameter keeps getting bigger as time moves on (which I find to be personally relatable).

When the big bang occurred, the universe was a writhing hellfire—a cauldron of enormous heat, choked with radiation and elementary particles. After matter and antimatter annihilated each other, a soup of the remaining particle matter and electromagnetic radiation hurtled outward, likely propelled by dark energy. This expansion of our domain continues today, almost fourteen billion years later.

So, for nearly fourteen billion years, electromagnetic particles have been zooming across the cosmos at the speed of light, embedded within a spacetime fabric that is also expanding. What I'm trying to say is there's a whole lot of universe out there. And we cannot see it all.

What we *can* measure is a spherical area of the universe around us (called the "observable universe") with a radius of approximately forty-six billion light-years. Now, this may not make sense to those of you with keen math skills. How can particles traveling at the speed of light end up at a distance that is farther than their finite speed could have taken them? Well, it's kind of like those people movers at large international airports: two people could be walking at the same relative pace, but if one is on the mover, their velocity is increased proportionally with the speed of the moving walkway itself. Similarly, our expansion of spacetime serves to move light and matter at distances further than their individual velocities alone could dictate.

There has been a large range of estimates for how many galaxies might be contained within this observable section of the universe: between one hundred billion to two trillion! But this is only part of the answer you're looking for—in fact, you asked me how many galaxies are in the *entire* universe. That, dear reader, is dependent upon one critical fact: does the universe have a finite border, or is it infinite in size?

At this point in human history, the true size of the universe is a bit speculative—our instrumentation can't detect anything beyond the boundary of what is observable, nor do we think it will ever be able to. So we aren't certain if the universe is a few times larger than what we can see at this point, or *incalculably* larger. Correspondingly, the number of total galaxies could then range from a few hundred billion to . . . well . . . a billion multiplied by infinity.

So, how many galaxies are in the universe? A hell of a lot.

IF THERE ARE TRILLIONS AND TRILLIONS OF STARS WITH POTENTIAL PLANETS, WHY HAVEN'T WE DISCOVERED ALIENS YET?

As a scientist, and also as someone who backpacks, I think of this situation often:

Me, alone in the woods with a flashlight, late to set up my campsite. As I peer into the forested blackness, I'm suddenly bathed in a beam of blinding blue light. Beings from another world descend to the ground before me, and I have the chance to make an historic point of contact with an extra-terrestrial civilization.

I'm overwhelmed with the magnitude of this responsibility—me, the representative for *all* of humanity. I slowly lower my backpack, and I ready myself to ask poignant, objective questions about their species, their mode of travel, and the nature of their planet of origin. I open my mouth to utter my first verbal communication to them, and I manage to say:

"Holy shiiiiiiiiiiiiiiiiii . . . !!!!!!!!"

I'm just being honest with you. And with myself.

Back to the question at hand: The chances of me absolutely blowing it in front of celestial travelers is quite slim. For a few reasons.

First, the universe is a big, *big* place with expansive distances of separation between stars and from galaxy to galaxy—we're talking light-years of travel. This means if we were to discover someone else, it would have to be on our turf. We just don't have the ability to reach even our closest neighbors. So their technological capabilities and understanding of space travel would have to be far more advanced than ours . . . by a long shot. Bottom line: someone else would have to chance upon this teensy blue marble and be intrigued enough to come and knock on our door.

Second, our idea of what is "alive" or "intelligent" is unbelievably geocentric. In other words, we only know how living things work on our planet: carbon-based species with the capacity to reproduce and grow on their own—the ones we classify as "intelligent" push air through their mouth-holes and make coordinated noises that culminate in language. So when we look for life, we look for things that we understand to be true about life on our *own* planet. But that classification is astoundingly narrow. Life doesn't have to be carbon-based. At all. Additionally, extraterrestrial life could look completely inert to us (think a matte-gray blob, piled up on the ground), but could still communicate telepathically without us ever having any inkling. It's possible! That being said, it's damn hard to find something if you have no idea what you're looking for. So even if our definition of life is incredibly specific, it's the only one we've got. Unfortunately, that may lead us to misclassification issues in the future.

Third, we don't even have a good grasp on the mathematics of how many intelligent civilizations may be out there. Our planet has only been around for approximately four billion years, and *it* has managed to produce intelligent life. But the universe itself is about fourteen billion years old, so there are plenty of planetary systems that are much older than we are. Surely, some of them have sprouted life *at least* as advanced as us . . . right? In the 1960s, a scientist named Frank Drake tried to quantify what this number might be. He came up with the Drake equation (named after himself, obviously—scientists love that), which was meant to estimate how many extraterrestrial civilizations may exist that could theoretically communicate with us. The Drake equation goes something like this:

> [# of intelligent extraterrestrial civilizations] = [average rate of star formation] * [proportion of stars with planets] * [average number of planets with the ability to harbor life] * [proportion of planets that *actually* develop life] * [proportion of planets with intelligent life] * [proportion of planets with intelligent life that transmit some kind of signal that we could detect] * [how long intelligent life has been releasing these signals]

The problem with this equation is that we don't have a good estimate for most of these variables. The average rate of star formation, and to a certain extent, the proportion of stars with planets, can be estimated with relative confidence. But if the scientific community is being honest with itself? Applying a number to the rest of the variables is a gigantic shot in the dark. The fact is, at this point, we can't even confirm or deny the existence of other life *within our solar system*, to say nothing of planets outside of it. Therefore, trying to estimate how many planets on average develop intelligent life that pumps out detectable electromagnetic frequencies? I dunno . . . pick your favorite number, I guess.

To summarize: while there are trillions and trillions of stars that could possibly offer up an extraterrestrial civilization, we simply haven't found any yet due to our limited technological capabilities, the immense distances between us and those candidate stars, our paltry definition of life itself, and no good heading for us to currently follow.

But, ya know, happy hunting!

WHY IS EARTH THE ONLY PLANET IN OUR SOLAR SYSTEM THAT HARBORS LIFE?

Great question, but it's peppered with a dash of wording bias. To clarify, we don't know if we are the only place in the solar system that brandishes a little population of wriggling entities. *Several* orbiting bodies within our solar system actually represent possible candidates for housing life. So let's attack this in two parts.

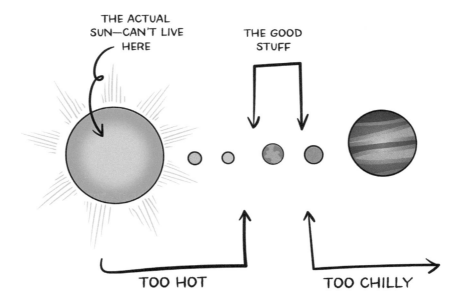

PART I: WE AIN'T THE ONLY GIG IN TOWN

Europa, Enceladus, Mars, Io, Titan (moon, moon, planet, moon, moon, respectively)—these five orbiting bodies are tantalizing to astrobiologists for their potentially life-sustaining environments. While surface conditions on many of these little worlds offer bleak hope for nurturing life, they may provide subsurface refuge.

Let's take a peek at Europa. Europa is a moon of Jupiter, and if you were doing a flyby, you might describe it as a big snowball. However, beneath its crusty snow-cone surface is an immense ocean—possibly ten times deeper (or more!) than the deepest part of any Earth ocean. Scientists have observed telescopically that the surface is made from shifting water ice, an indication that there is liquid not far below it, and that the liquid is likely also water. More still, there are breaks in Europa's chemical poker face, revealing irradiated deposits of sodium chloride on its surface. Taken together, what does this all mean? Well, Europa may be hiding one massive salty ocean, which sounds strikingly similar to the ones on our planet. If we're lucky, Europa could be home to an ocean of microbes. If we're *very* lucky, Europa might be home to a heaving ice kraken.

Up next: Titan. Titan is a moon of Saturn, pickled in a choking fog of nitrogenous gas and organic compounds. The surface has a temperature of a balmy −300°F, and it is likely devoid of liquid surface water. But! There may be liquid in the form of methane- and ethane-filled lakes. Between the chilly temperatures, soupy gases, and parched landscape, earthlike life would not find this to be a cozy home. However, scientists theorize that life *could* exist in a form much different from ours, taking shelter within the stagnant methane bogs. In fact, multiple lab-based experiments have been conducted to assess the formation of critical prelife molecules, based on the raw materials available on Titan. The wacky findings: despite being made of inherently different molecules from ours, scientists discovered that Titan molecules might behave similarly to our own, coalescing to form complex structures necessary to kick-start the evolution of life.

PART II: EARTH IS THE FAVORITE CHILD

Now that we know life might exist elsewhere in our celestial lineup, let's talk about why Earth happens to be *the* planet. You know what I'm talking about: the planet that spawned the species that wrote "Moonlight Sonata," that built rockets and found a way to escape its planet's gravity, that started the cinnamon challenge. Much of our special seat in the cosmos comes down to location—we are, more or less, a perfect distance away from the sun. Our location ensures we are close enough to the sun's heat to maintain the availability of liquid water, but far enough away that the water doesn't evaporate completely. Additionally, because of Earth's stages of solid and liquid core, it generates a magnetosphere that acts as a protective barrier to keep solar winds from stripping the planet's atmosphere away. Finally, the planet is rich in carbon, nitrogen, oxygen, and hydrogen, which served as brilliant building blocks for molecules essential to life's humble origin story.

So, as special as we may feel, we're the only planet that harbors life for two reasons: (1) our current ignorance of what else is actually out there and (2) winning one hell of a location lottery.

Some planets have all the luck.

WHY DO STARS FLICKER IN THE NIGHT SKY?

Many of the things our parents told us were untrue: "If you keep making that face, it will get stuck that way." Some of the things our parents told us were spot-on: "If it's flickering, it's a star; if it's not flickering it's a planet." (They also mentioned something about hairy palms that I can't seem to remember . . .)

The light from stars appears to throb and twinkle because they are incredibly far away from us. From our vantage point, the origin of starlight looks like a small illuminated pinhole in space (even though it comes from a furious ball of hellfire). These sparse photons refract and become slightly scattered as they pass through Earth's atmosphere. The bending of these rather small beams of light disrupts the steady flow of photons perceived by the human eye, and therefore, the pinpoints appear to flicker.

Planets are much closer. As such, their radius in the sky is a bit wider and so, too, is the swath of light they project. So while planetary photons (which are actually reflected photons from the sun) are still subject to similar refraction patterns, the bigger beam compensates for errant movement among photons, and we perceive this to be a steady light source.

When I was a small child, I was under the impression that stars flickered because they were candles suspended in space. I assumed that the twinkling illusion was from the outer-space disturbance of the small flames.

And this, dear reader, is precisely why you are not allowed to vote until you're eighteen years of age.

SCIENTISTS HAVE YET TO DISPROVE THAT STARS FLICKER BECAUSE THEY THROW SPACE RAVES

ACKNOWLEDGMENTS

I've come to find that it's not grand cosmic occurrences that irrevocably alter the course of our individual lives. It's trivial moments—the met glance from a stranger, an uncommon street you chose to travel home by, or maybe the fleeting notion that you should write a science reference book for adults, which will also be illustrated in crayon.

The meanderings of my mind, during one perfectly insignificant Tuesday morning commute, ultimately led me to rediscover my passion for human connection through the written word. But I would be remiss if I didn't credit this project (and my burgeoning career as an author) to two exceptionally important enablers.

To my fellow, heavily tattooed scientist and author Gareth Worthington:

Thank you for making a call from Switzerland to listen to me breathily pace around the hospital parking lot, railing on about the merits of bizarre crayon depictions of science. You never flinched, you never chided me, and you gave me the unmitigated gall to believe in this work. Your enthusiasm and support were integral to making this happen. "Thank you" isn't enough, dear friend. But please let it act as a temporary surrogate until I can hug you properly.

To my FBB, my "pretty girl with long hair—woo!" and literary co-conspirator, Renée Fountain:

I'm not entirely sure what you saw in me, poorly prepared and huddled in the grainy mise-en-scène of an outdated web camera. But I'm glad you saw it. Thank you for always being there to talk about life and literature. Thank you for always making time to field both my good and shitty ideas (and for telling me, with absolute conviction, that they are both good and shitty). This journey has been made possible by your guiding hand and your continued words of wisdom. I cannot wait to see where we go together. I certainly hope that you've found a place for Ziggy.

NOTES

1   Barbara O. Rennard et al., "Chicken Soup Inhibits Neutrophil Chemotaxis *In Vitro*," *Chest* 118, no. 4 (October 1, 2000): 1150–57, https://doi.org/10.1378/chest.118.4.1150.

2   Charles Darwin, *The Power of Movement in Plants*, assisted by Francis Darwin (London: John Murray, 1880).

3   Vidya Chivukula and Shivaraman Ramaswamy, "Effect of Different Types of Music on Rosa Chinensis Plants," *International Journal of Environmental Science and Development* 5, no. 5 (October 2014): 431–34, https://doi.org/10.7763/ijesd.2014.v5.522; Reda H. E. Hassanien et al., "Advances in Effects of Sound Waves on Plants," *Journal of Integrative Agriculture* 13, no. 2 (February 2014): 335–48, https://doi.org/10.1016/s2095-3119(13)60492-x; Md. Emran Khan Chowdhury, Hyoun-Sub Lim, and Hanhong Bae, "Update on the Effects of Sound Wave on Plants," *Research in Plant Disease* 20, no. 1 (2014): 1–7, https://doi.org/10.5423/rpd.2014.20.1.001.

4   David P. Fernandez, Daria J. Kuss, and Mark D. Griffiths, "The Pornography 'Rebooting' Experience: A Qualitative Analysis of Abstinence Journals on an Online Pornography Abstinence Forum," *Archives of Sexual Behavior* 50 (2021): 711–28, https://doi.org/10.1007/s10508-020-01858-w.

5   Marta Kramkowska, Teresa Grzelak, and Krystyna Czyżewska, "Benefits and Risks Associated with Genetically Modified Food Products," *Annals of Agricultural and Environmental Medicine* 20, no. 3 (September 2013): 413–419, https://pubmed.ncbi.nlm.nih.gov/24069841/; Artemis Dona and Ioannis S. Arvanitoyannis, "Health Risks of Genetically Modified Foods," *Critical Reviews in Food Science and Nutrition* 49, no. 2 (2009): 164–75, https://doi.org/10.1080/10408390701855993.

6   F. Belva et al., "Chromosomal Abnormalities after ICSI in Relation to Semen Parameters: Results in 1114 Fetuses and 1391 Neonates from a Single Center," *Human Reproduction* 35, no. 9 (September 2020): 2149–62, https://doi.org/10.1093/humrep/deaa162; Gerald Lawson and Richard Fletcher, "Delayed Fatherhood," *Journal of Family Planning and Reproductive Health Care* 40, no. 4 (September 19, 2014): 283–88, https://doi.org/10.1136/jfprhc-2013-100866; Alexander N. Yatsenko and Paul J. Turek, "Reproductive Genetics and the Aging Male," *Journal of Assisted Reproduction and Genetics* 35 (June 2018): 933–41, https://doi.org/10.1007/s10815-018-1148-y.

7   Peter Daszak et al., "Workshop Report on Biodiversity and Pandemics of the Intergovernmental Platform on Biodiversity and Ecosystem Services (IPBES)," Intergovernmental Science-Policy Platform on Biodiversity and Ecosystem Services (October 29, 2020), http://dx.doi.org/10.5281/zenodo.4147317.

8   Antonia M. Calafat et al., "Urinary Concentrations of Bisphenol A and 4-Nonylphenol in a Human Reference Population," *Environmental Health Perspectives* 113, no. 4 (April 1, 2005): 391–95, https://doi.org/10.1289/ehp.7534; Centers for Disease Control and Prevention, "Urinary Bisphenol A (2003–2010)," National Report on Human Exposure to Environmental Chemicals, accessed December 7, 2022, https://www.cdc.gov/exposurereport/report/pdf/cgroup31_URXBPH_1999ua-p.pdf.

9   Claudia Pivonello et al., "Bisphenol A: An Emerging Threat to Female Fertility," *Reproductive Biology and Endocrinology* 18, no. 22 (March 14, 2020), https://doi.org/10.1186/s12958-019-0558-8; De-Kun Li et al., "Relationship between Urine Bisphenol-A Level and Declining Male Sexual Function," *Journal of Andrology* 31 (January 2, 2013): 500–506, https://doi.org/10.2164/jandrol.110.010413; Maohua Miao et al.,"*In Utero* Exposure to Bisphenol-A and Its Effect on Birth Weight of Offspring," *Reproductive Toxicology* 32, no. 1 (July 2011): 64–68, https://doi.org/10.1016/j.reprotox.2011.03.002; Maede Ejaredar et al., "Bisphenol A Exposure and Children's Behavior: A Systematic Review," *Journal of Exposure Science & Environmental Epidemiology* 27 (2017): 175–83, https://doi.org/10.1038/jes.2016.8; Marcelino Pérez-Bermejo, Irene Mas-Pérez, and Maria Teresa Murillo-Llorente, "The Role of the Bisphenol A in Diabetes and Obesity," *Biomedicines* 9, no. 6 (June 2021): 666, https://doi.org/10.3390/biomedicines9060666; Xiaoqian Gao and Hong-Sheng Wang, "Impact of Bisphenol A on the Cardiovascular System—Epidemiological and Experimental Evidence and Molecular Mechanisms," *International Journal of Environmental Research and Public Health* 11, no. 8 (August 15, 2014): 8399–8413, https://doi.org/10.3390/ijerph110808399.

10   Elie Dolgin, "Colour Blindness Corrected by Gene Therapy," *Nature* (2009), https://doi.org/10.1038/news.2009.921.

11   M. Riebe et al. "Deterministic Quantum Teleportation with Atoms," *Nature* 429, (June 17, 2004): 734–37, https://doi.org/10.1038/nature02570; John Boviatsis and Evangelos Voutsinas, "Quantum Control and Entanglement of Two Electrons in a Double Quantum Dot Structure," AIP Conference Proceedings 963 (2007): 740–743, https://doi.org/10.1063/1.2836196; S. Haroche, "Engineering Entanglement between Atoms and Photons in a Cavity," *Quantum Coherence and Decoherence* (1999): 13–18, https://doi.org/10.1016/b978-044450091-5/50006-3; C. Marletto et al., "Entanglement between Living Bacteria and Quantized Light Witnessed by Rabi Splitting," *Journal of Physics Communications* 2, no. 10 (October 10, 2018): 101001, https://doi.org/10.1088/2399-6528/aae224.

12   Julia Geynisman-Tan et al., "Bare versus Hair: Do Pubic Hair Grooming Preferences Dictate the Urogenital Microbiome?" *Female Pelvic Medicine & Reconstructive Surgery* 27, no. 9 (September 2021): 532–37, https://doi.org/10.1097/spv.0000000000000968.

13   Sanela Hadžić, Ismir Kukić, and Jasmin Zvorničanin, "The Prevalence of Eyelid Myokymia in Medical Students," *British Journal of Medicine and Medical Research* 14, no. 6 (March 2016): 1–6, https://doi.org/10.9734/bjmmr/2016/24910; Rudrani Banik and Neil R. Miller, "Chronic Myokymia Limited to the Eyelid Is a Benign Condition," *Journal of Neuro-Ophthalmology* 24, no. 4 (December 2004): 290–92, https://doi.org/10.1097/00041327-200412000-00003.

14   Abhi Humar, "How Live Liver Transplants Could Save Thousands of Lives," The Conversation, April 25, 2018, https://theconversation.com/how-live-liver-transplants-could-save-thousands-of-lives-94698.

15   Andrea Cavallo et al., "Decoding Intentions from Movement Kinematics," *Scientific Reports* 6, no. 37036 (2016), https://doi.org/10.1038/srep37036.

16   H. P. T. Ammon, "Biochemical Mechanism of Caffeine Tolerance," *Archiv der Pharmazie* 324, no. 5 (1991): 261–67, https://doi.org/10.1002/ardp.19913240502; Per Svenningsson, George G. Nomikos, and Bertil B. Fredholm, "The Stimulatory Action and the Development of Tolerance to Caffeine Is Associated with Alterations in Gene Expression in Specific Brain Regions," *Journal of Neuroscience* 19, no. 10 (May 15, 1999): 4011–22, https://doi.org/10.1523/jneurosci.19-10-04011.1999; Chyan E. Lau and John L. Falk, "Dose-Dependent Surmountability of Locomotor Activity in Caffeine Tolerance," *Pharmacology Biochemistry and Behavior* 52, no. 1 (September 1995): 139–43, https://doi.org/10.1016/0091-3057(95)00066-6.

17   Kazunori Akiyama et al., Event Horizon Telescope Collaboration, "First M87 Event Horizon Telescope Results. I. The Shadow of the Supermassive Black Hole," *Astrophysical Journal Letters*, 875, no. 1 (April 10, 2019), https://doi.org/10.3847/2041-8213/ab0ec7.